PHYSICAL SCIENCE FOR BIOLOGISTS

Physical Sciences

Editor
R. O. DAVIES
D.PHIL, M.SC.
Professor of Physics, University College, Cardiff

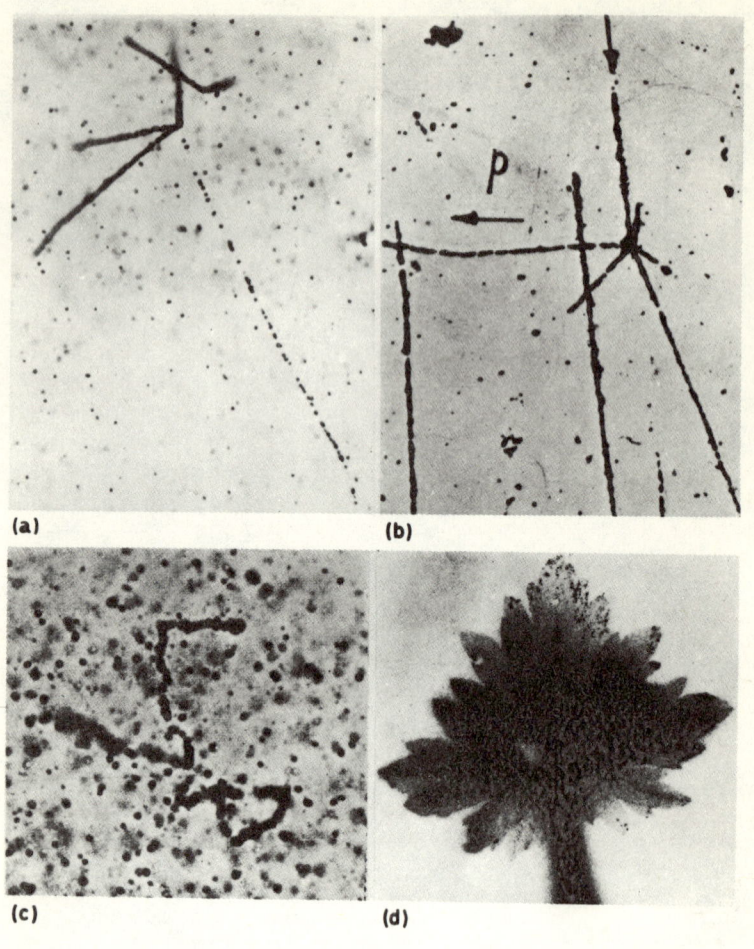

The tracks of various particles in photographic film (see p. 189). (a) Short dense tracks of α-particles given off by thorium and its decay products. The lighter, longer track is caused by a β-particle from one of the daughter nuclei (Bi^{212}). (b) A nuclear reaction. A He^3 nucleus strikes a nitrogen nucleus, yielding four α-particles and a proton (track p). (c) Track of a slow β-particle. (d) Autoradiograph of a buttercup leaf which has absorbed phosphate containing radioactive P^{32}. The individual β-tracks are not resolved. (a), (b) and (c) are magnified approximately × 500; (d) about × 3.

Reproduced from 'Applications of Nuclear Physics' by kind permission of Professor J.H. Fremlin and The English Universities Press Ltd

PHYSICAL SCIENCE
FOR BIOLOGISTS

J. A. Edgington
Lecturer in Physics, Queen Mary College
University of London

&

H. J. Sherman
Principal Scientific Officer
Daresbury Nuclear Physics Laboratory

HUTCHINSON UNIVERSITY LIBRARY
LONDON

HUTCHINSON & CO (*Publishers*) LTD
3 Fitzroy Square, London W1

London Melbourne Sydney Auckland
Wellington Johannesburg Cape Town
and agencies throughout the world

First published 1971

The paperback edition of this book is sold subject to the condition that it shall not, by way of trade or otherwise, be lent, re-sold, hired out or otherwise circulated without the publisher's prior consent in any form of binding or cover other than that in which it is published and without a similar condition being imposed on the subsequent purchaser

The cover of the paperback edition shows an X-ray diffraction pattern from a fibre of sodium deoxyribonucleic acid at 92% relative humidity. By courtesy of the Medical Research Council Biophysics Unit, King's College, London.

© J.A. Edgington and H.J. Sherman 1971

This book has been set in cold type by E.W.C. Wilkins & Associates Ltd, London, printed in Great Britain by Anchor Press, and bound by Wm. Brendon, both of Tiptree, Essex

ISBN 0 09 107860 1 (cased).
0 09 107861 X (paper).

CONTENTS

Introduction   vii

Units and numerical constants   xi

1 Mechanics   1
- Vectors   1
- Kinematics   5
- The kinematics of circular motion   6
- Dynamics: Newton's laws of motion   10
- Newton's law of gravitation   15
- Work and power   17
- Energy   20
- Conservation of energy   24
- Vibrations and waves   27
- Problem-solving in mechanics   34

2 The Molecular Nature of Matter   37
- Kinetic theory of gases   37
- Brownian motion   42
- The mean free path and the size of molecules   45
- Diffusion   47
- Osmosis   52
- Sedimentation and the distribution of molecules in space   57
- The Maxwell distribution   63
- Viscosity   68

3 Thermodynamics   73
- The first law of thermodynamics   73
- Enthalpy   76
- Entropy   80
- The second law of thermodynamics   81
- Gibbs' free energy: its use in reaction analysis   84

4 Electrochemistry — 89

- The flow of electricity in metals — 89
- Electrolytic conduction — 91
- Ionic velocity and transport number — 94
- Electrochemical cells — 98
- Concentration cells and *pH* values — 103

5 Optics and the Wave Nature of Light — 109

- The nature of light — 109
- Geometrical optics: the design of optical instruments — 112
- Interference — 117
- Diffraction — 124
- Resolving power: the limit set by diffraction — 130
- The polarisation of light — 136
- Laser light — 143

6 Modern Physics: Particles, Waves and Probabilities — 147

- Gas discharges and atomic structure — 149
- The photoelectric effect — 151
- The production and diffraction of X-rays — 153
- The wave nature of electrons — 156
- Quantization and atomic spectra — 160
- What are matter waves? — 164
- Further quantum numbers — 166
- The solid state — 168

7 Nuclear Radiations — 173

- The atomic nucleus — 174
- Radioactivity — 178
- The interaction of radiation with matter — 185
- Detection of radiation — 191
- Biological effects of radiation — 197
- Radiation dosimetry — 198
- Beneficial uses of radiation — 203

Exercises and Answers — 207

Bibliography — 219

Index — 221

INTRODUCTION

Textbooks on general physics usually fall into one of two categories. The first is designed for students who have studied physics and mathematics up to Advanced level standard at school; its purpose is to contain within two covers all the physics required for the first year or so of a university course in physics or physical sciences. Consequently, as many topics as possible are covered, and since the readers are usually taking additional mathematics courses the level of mathematical proficiency assumed is quite high. The second class of textbooks is intended for arts students who have not studied science to Advanced level at school but who are either changing to physics at university or else require such a course as part of a so-called 'foundation year'. Such books avoid relatively difficult mathematics but in most other respects are similar to the first category. They attempt to cover much the same ground, and the approach to the subject is still that of the professional physicist, albeit often enlivened by a chatty style and a wealth of examples drawn from everyday experience.

We feel there is an unmet demand for a third category of textbook, written for students who have studied biological sciences at school, probably at the expense of mathematics beyond Ordinary level standard. Such students find when they reach university that some knowledge of physical concepts and techniques is necessary for many of the courses they wish to take. In particular, medical students at the pre-clinical stage of their studies, and biology students in science faculties offering a number of interdisciplinary options, are likely to find such a textbook useful. Students of similar subjects outside universities, such as radiography, nursing and ophthalmics are also likely to benefit from such a book.

The present book is based on our experience in teaching biology students at Queen Mary College during the first four years of the London University 'New Degree' regulations, under which courses in the Faculty of Science are open to all students, regardless of their department of registration. We found a large, and steadily increasing, demand from the various departments of biological science for their students to receive some basic physics in their first year, specifically to prepare them for later biology courses. After initial discussions with biology teachers we drew up a syllabus for such a course. In subsequent years this was modified greatly until it reached the stage presented here. Though many improvements still suggest themselves, we feel that the subjects discussed, and their method of presentation, form a sufficiently precise framework to be worth putting into print.

The main features of this book are as follows. First, we have made no attempt to cover *all* of so-called 'elementary' physics, much of which is either too recondite, or else inappropriate to the needs of biology students (and sometimes both). Thus the reader will find no discussion of magnetism, rotational dynamics, relativity, electrostatics or sub-nuclear physics. On the other hand we have included some topics which are not generally taken to be the concern of university physics courses at all, but which are assumed to be known to the biology student by his third year! The chapter on electrochemistry is included for this reason, and also that section of the last chapter dealing with health physics. One aspect of this approach is that sub-headings whose names may seem familiar to those taking Advanced level physics (such as diffusion, viscosity, and thermodynamics) are treated very differently, though no less extensively, than they would be in such a course. Thus this book is by no means a substitute for an Advanced level text — the topics covered hardly overlap at all.

The second feature lies in our extensive use of examples, which are where possible drawn from the life sciences. Whenever a numerical result is obtained we have attempted to illustrate its practical significance in this way. Similarly, having derived an algebraic relation we illustrate it with a (hopefully) realistic example, so that the student realises the kind of situations and problems that can be tackled using that approach. Such illustrative material is indicated by the word **Example**. We have *not*, however, attempted to write a textbook of biophysics. Wherever the biological applications loom large the reader is referred to the bibliography for further reading.

Introduction

Finally, the book uses no advanced mathematics. Calculus is not used, and though we have had to indulge in occasional mental contortions to achieve this aim, we feel that the resultant stress on the qualitative aspects should help all readers, not only the less mathematically able ones. Many of our examples and calculations use an order-of-magnitude approach, to try to instil the attitude that a crude feeling for the *size* of a measurable quantity is at least as important as a knowledge of its exact value. There seems to be too little of this attitude amongst physicists, certainly at schools and sometimes also in universities, and we would guess that the tedious arithmetic drill which is still too often encouraged must have turned many biology students away from physics at school.

This, therefore, is a textbook for students with a reasonably mature scientific outlook but little recent acquaintance with physics or mathematics. The order of the subject-matter reflects this. Chapter 1, on mechanics, could perhaps be omitted by students who have studied some applied mathematics at school, though even they should attempt the appropriate exercises at the end of the book. Chapter 2 uses these mechanical laws, plus the molecular idea of matter, to derive many of the macroscopic properties of real matter, particularly fluids. The law of conservation of energy is further expanded in Chapters 3 and 4 to deduce some results of thermodynamics and to apply these to electrochemistry. In Chapter 5 we turn to wave motion, and the laws of physical optics are applied to a variety of actual situations. The final two chapters constitute, respectively, an introduction to the ideas and applications of quantum theory, and a summary of the main results of nuclear radiation physics.

One last point should be made. By itself, this textbook is incomplete. It lacks laboratory practice in applying the results. The course which we have devised at Queen Mary College has associated with it a laboratory with a relatively small number of experiments specially designed to illustrate the close relationship between physical facts, experimental techniques and actual laboratory or fieldwork apparatus. For example, students use centrifuges to show the character of centrifugal forces, to illustrate the laws and techniques of sedimentation, and finally to apply these to a particular problem of sediment separation. Such laboratory work should always accompany any course based on this book.

J.A.E.

H.J.S.

UNITS AND NUMERICAL CONSTANTS

Throughout the book we use the international (SI) system of units, except when common usage sometimes makes another system more appropriate. The basic units are the metre (m), second (s), kilogram (kg), degree kelvin (°K), and coulomb (C). All the derived units, such as those of power and energy, are obtained by multiplying together the appropriate basic units. Thus the unit of momentum, defined as mass times velocity, is (kg m s^{-1}). Some of the derived units are given special names, but the only ones we shall use are the following.

Mechanical units:
  Force – the unit is the newton (N). $1\,\text{N} = 1\,\text{kg m s}^{-2}$
  Work and *energy* – the unit is the joule (J). $1\,\text{J} = 1\,\text{kg m}^2\,\text{s}^{-2}$
  $\phantom{Work\ and\ energy\ -\ the\ unit\ is\ the\ joule\ (J).\ 1\,\text{J}} = 1\,\text{N m}$
  Power – the unit is the watt (W). $1\,\text{W} = 1\,\text{kg m}^2\,\text{s}^{-3}$
  $\phantom{Power\ -\ the\ unit\ is\ the\ watt\ (W).\ 1\,\text{W}} = 1\,\text{J s}^{-1}$
  Frequency – the unit is the hertz (Hz). $1\,\text{Hz} = 1$ cycle per second

Electrical units:
  The unit of *electric current* is the ampere (A). $1\,\text{A} = 1\,\text{C s}^{-1}$
  The unit of *electric potential* is the volt (V), defined by stating that a current of 1 A flowing through a potential difference of 1 V yields a power output of 1 W.
  The unit of *resistance* is the ohm (Ω), defined by stating that a current of 1 A flowing through a resistance of 1 Ω will produce a potential difference of 1 V across the resistance.

The following exceptions to the SI system are sometimes used:
1 *litre* = 10^{-3} m^3
1 *atmosphere* ≈ 10^5 N m^{-2} (≈ pressure due to 760 mm of mercury)
A *normal* solution contains one one gram-equivalent of solute in
   1 litre of solution
0°C = 273·16°K
1 *calorie* = 4·18 J
1 *electron volt* (eV) = 1·602 × 10^{-19} J

In the SI system, large and small quantities are usually expressed in powers of 10 of the basic units. Every third power of 10 (that is, 10^3, 10^6, 10^9 etc) has a special name which is used as a prefix to the unit name. Thus 10^{-3} m = 1 *milli*metre. Some of the prefixes and their values are listed below.

milli-	(10^{-3})	kilo-	(10^3)
micro-	(10^{-6})	mega-	(10^6)
nano-	(10^{-9})	giga-	(10^9)

The following list contains all the numerical constants that are used in the book. All values are given in SI units.

the gravitational constant $G = 6·67 \times 10^{-11}$ m^3 kg^{-1} s^{-2}
acceleration due to gravity at the earth's surface $g = 9·81$ m s^{-1}
Avogadro's number $N_0 = 6·02 \times 10^{23}$ molecules per mole
Boltzmann's constant $k = 1·38 \times 10^{-23}$ J °K^{-1}
universal gas constant $R = 8·31$ J °K^{-1} mol^{-1}
                             $= 1·99$ calories °K^{-1} mol^{-1}
mass of an electron $m_e = 9·11 \times 10^{-31}$ kg
mass of a proton $m_p = 1·67 \times 10^{-27}$ kg
charge of an electron (proton) $e = 1·602 \times 10^{-19}$ C
the Faraday (charge on a gram mole of a univalent ion)
                             $F = 96 487$ C
velocity of light $c = 2·998 \times 10^8$ m s^{-1}
Planck's constant $h = 6·63 \times 10^{-34}$ J s
Rydberg's constant $Ry = 2·18 \times 10^{-18}$ J
                             $= 13·6$ eV
the curie (Ci) is a disintegration rate of $3·7 \times 10^{10}$ per second

I

MECHANICS

Mechanics is the study of how and why objects move; or, on those occasions when they are at rest, why they do *not* move. These two aspects of motion, its presence and its absence, account for the elementary school division into statics and dynamics. Since the former, however, is only a special case of the latter we shall not stress the difference but shall concentrate on movement as such; how to describe it, what causes it and what laws it obeys. Specifically, in the first section we develop a notation (use of vectors) which we use in later sections to describe how bodies move (kinematics) and why they move (dynamics). This leads us to Newton's laws of motion and we show how they can be applied, as they were by Newton himself, to explain the motion of bodies under the influence of gravity. In later sections the commonplace ideas of work, power, and energy are assigned a rigorous meaning, and conservation of energy (perhaps the most important single concept in the whole book) is introduced for the first time. Finally these ideas are extended to include vibrational and wave motion.

Vectors

Consideration of many physical properties such as volume, temperature, mass, electric charge and so on shows that they are completely determined by their magnitude, that is by a single number. Only one number need be given to specify a temperature, as 37°C, or the time, as 12.30 p.m. Such quantities are called *scalars*.

Many other physical properties are not completely determined by a single magnitude. An example is the displacement of a particle from one point to another on a table top. To convey all the information about the movement of the particle it is necessary to state, not only the distance moved (*magnitude*) but also the *direction* of the movement. Such quantities, which require a direction as well as a magnitude for their specification, are called *vectors*. Examples are velocity, acceleration and force. For example, a high jumper and a long jumper may leap with the same effort (magnitude of the force), but because this effort is applied in different directions the results differ considerably. Our problem is to express this difference between vectors in a simple, explicit manner.

Various notations are used in science as a form of shorthand. For example, the equation $y = mx + c$ is a shorthand way of describing a certain relation between pairs of x and y values. Alternatively we could use a graphical notation, drawing X and Y axes at right angles and representing this same relation by a straight line drawn on the graph (Fig. 1.1a). A similar graphical notation has been developed for use with vectors, and we will now outline its main features.

A displacement from the origin O to the point A (Fig. 1.1b) is represented graphically as a vector by a line having the same direction as the displacement and whose length is equal to the displacement. The direction of the displacement is indicated by an arrow as shown in Fig. 1.1(b). The algebraic notation for a vector is a capital letter underlined, as $\underline{V}$, a bold face letter as **V**, or often a letter with an arrow over it, as $\vec{V}$. Numerically a pair of numbers are required to specify the vector. There are many different ways of doing this. For example, the vector shown in Fig. 1.1(b) could be specified by saying that its magnitude was $\sqrt{13}$ and that it made a positive angle of 33·7° with the X axis.

Vectors can be added to each other, giving as a result another vector. Consider a particle displaced from the origin O to point B and let this displacement be represented by a vector $\mathbf{V_1}$. Then consider the particle to be displaced from point B to point C and let that displacement be represented by a vector $\mathbf{V_2}$. It is clear that the total displacement can be represented by a vector $\mathbf{V_3}$ which goes from the origin to point C (see Fig. 1.1c). Thus the result of the two displacements is equivalent to a single displacement. This is written algebraically as

$$\mathbf{V_3} = \mathbf{V_1} + \mathbf{V_2} \tag{1.1}$$

Mechanics

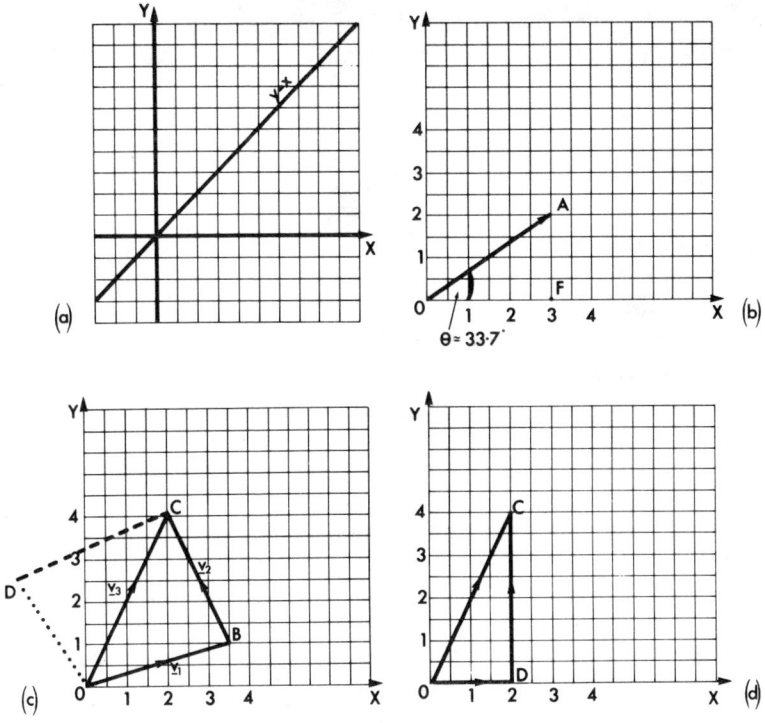

Fig. 1.1
(a) Graphical representation of the straight line $y = x$.
(b) Two possible ways of expressing the displacement OA.
(c) A vector displacement is the sum of two vectors.
(d) The components of the vector OC of the previous diagram.

Since vectors are being added it is important to write the quantities as vectors in eqn (1.1). Intuitively one can see that the same resultant displacement would have been obtained if displacement V_2 had been performed first (shown as the dotted line OD) followed by displacement V_1 (shown as the dashed line DC). Algebraically this means that

$$V_3 = V_2 + V_1 \qquad (1.2)$$

Thus the addition of vectors, like that of ordinary scalars, is indifferent to the order in which the addition is done.

Brief consideration of Fig. 1.1(c) shows that there are an infinite number of pairs of vectors like V_1 and V_2 which can be drawn so that their resultant is the vector V_3. Each pair of vectors which, when added, give the vector **V** are said to be *components* of **V**. The most commonly used components are vectors which are perpendicular to each other; for instance, one vector parallel to the X axis and the other parallel to the Y axis. These vectors are referred to as the X component and Y component respectively.

Example: What are the X and Y components of the vectors $\mathbf{V}_1$, $\mathbf{V}_2$ and $\mathbf{V}_3$ shown in Fig. 1.1(c)? The X and Y components of $\mathbf{V}_3$ (OC) are shown in Fig. 1.1(d). The X component has length 2 units and is in the positive X direction. The Y component has length 4 units and is in the positive Y direction. This can be written as $(+2, +4)$. Similarly $\mathbf{V}_1$ has X and Y components $(+3\frac{1}{2}, +1)$ and $\mathbf{V}_2$ has X and Y components $(-1\frac{1}{2}, +3)$. Notice that the minus sign on the X component of $\mathbf{V}_2$ indicates that it is pointing in the negative direction along the X axis.

One of the reasons for the common use of X and Y components is that any vector problem is then split into two simpler problems. As an example of this we can take the addition of two vectors. Clearly the sum of the X components of the two vectors must equal the X component of the resultant and similarly for the Y components. Adding X components is easy because numerical addition taking account of the sign is all that is required. Thus the resultant X component, if $\mathbf{V}_1$ and $\mathbf{V}_2$ are added, is $+3\frac{1}{2} - 1\frac{1}{2} = 2$ units. This is the X component of $\mathbf{V}_3$. Similarly for Y components the result of adding $\mathbf{V}_1$ to $\mathbf{V}_2$ is $+1 +3 = +4$, which is the Y component of $\mathbf{V}_3$ (see Fig. 1.1d).

The X and Y components of a vector can be easily calculated provided the length of the vector and its angle with the X axis are known. In Fig. 1.1(b) the X component of OA is side OF of the right angled triangle OAF.

$$OF = OA \cos \theta$$

where OA is the length of the vector OA. Similarly the Y component is AF where

$$AF = OA \sin \theta$$

Example: Raindrops are falling vertically at $15 \, \text{m s}^{-1}$. With what velocity will they strike the windscreen of a car driven at $72 \, \text{km hr}^{-1}$? The solution is got by taking X and Y axes along the horizontal and vertical directions respectively. The X component of the raindrops' velocity is then $72 \, \text{km hr}^{-1} = 20 \, \text{m s}^{-1}$ and the Y component

Mechanics

is 15 m s⁻¹. These components form two sides of a (3, 4, 5) right-angled triangle whose hypotenuse is 25 m s⁻¹ in magnitude, making an angle of $\tan^{-1}(15/20) \approx 37°$ with the horizontal.

In conclusion it should be said that although the example of a vector which has been used in this discussion was displacement the results are perfectly general. Also, although the above discussion has been in terms of a two-dimensional system, there is nothing new or difficult about extending the treatment to three dimensions. The ideas introduced are valid for all vectors in a coordinate system of any number of dimensions. The book by Carman (see p. 219) is useful for those wanting to know more about vectors.

Kinematics

Consider a particle moving along the path shown in Fig. 1.2. It is at point P_1 at time t_1 and its position is given by the vector $\mathbf{r}_1$ drawn from the origin of coordinates O. Later, at time t_2, the particle reaches P_2, where its position is given by vector $\mathbf{r}_2$. The *average velocity* v is given by the vector $(\mathbf{r}_2 - \mathbf{r}_1)$ which is the vector displacement from P_1 to P_2, divided by the transit time $(t_2 - t_1)$. Thus

$$\mathbf{v} = \frac{(\mathbf{r}_2 - \mathbf{r}_1)}{(t_2 - t_1)} \tag{1.3}$$

This average velocity is a vector whose magnitude is equal to the distance between P_1 and P_2 divided by the transit time and whose direction is that of the displacement vector from P_1 to P_2.

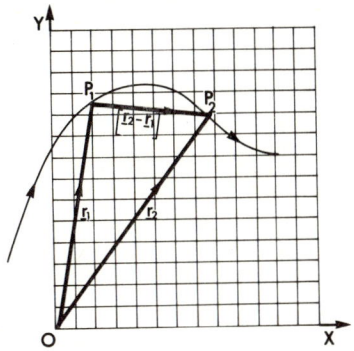

Fig. 1.2
Velocity and acceleration of a moving point.

Often, rather than an *average* velocity, it is more useful to talk of the actual velocity at each point. How can this be defined? In order to find out we consider the sequence of values of the quantity **v** given by eqn (1.3) as P_2 approaches P_1. The vector $\mathbf{r}_2$ in Fig. 1.2 is imagined to swing slowly round towards $\mathbf{r}_1$ as P_2 slides along the curve towards P_1 and at each new position the value of **v** is calculated from eqn (1.3). As P_2 gets very close to P_1, $(t_2 - t_1)$ will get very small. But so also will $(\mathbf{r}_2 - \mathbf{r}_1)$, and the value of **v** remains finite. The value of **v**, which is approached as P_2 approaches very close to P_1, is defined as the velocity at P_1. Like the average velocity it is a vector, having both magnitude and direction. From consideration of the geometry of Fig. 1.2 it can be seen that the direction of the velocity at P_1 is along the tangent to the curve at P_1. The appropriate unit for the magnitude of velocity (often called *speed*) is metre sec^{-1}.

The *acceleration* at point P_1 is defined in an analogous manner to this definition of velocity. Let the particle have velocity $\mathbf{v}_1$ at time t_1 when in position P_1 and velocity $\mathbf{v}_2$ at a later time t_2 when in position P_2. Consider the quantity ***a*** given by

$$\boldsymbol{a} = \frac{(\mathbf{v}_2 - \mathbf{v}_1)}{(t_2 - t_1)} \quad (1.4)$$

The value of ***a*** which is approached as P_2 moves towards P_1, is defined as the acceleration at P_1. As with velocity, acceleration is a vector quantity. The appropriate unit for the magnitude of acceleration is metre sec^{-2}.

Example: A car moves away from rest at traffic lights with an acceleration of 5 m s^{-2}, then travels at 20 m s^{-1} and finally decelerates again at 5 m s^{-2}, to stop at other traffic lights 200 m from the first. How long does the car take between traffic lights? From eqn (1.4) the time spent accelerating is 20/5 = 4 s (and a similar time is spent decelerating). During this time the average speed is 10 m s^{-1} so the car travels 40 m whilst accelerating (and also whilst decelerating). It thus travels 200 − (2 × 40) = 120 m at 20 m s^{-1}, taking 6 s to do so. The total time taken is therefore 4 + 6 + 4 = 14 seconds.

The kinematics of circular motion

Let us consider a point moving round a circle of radius r, centred at the origin O, with a constant speed v. Notice that the magnitude

Mechanics

of the velocity vector of the point (which we have called speed) is constant, but the direction of the vector is not. Thus, it would be incorrect to say that the point was moving round the circle with a constant (vector) velocity.

Let the distance from point P_0 on the X axis to P_1 be s metres, measured around the circle (Fig. 1.3). As the point moves from P_0 to P_1 the radius sweeps through an angle θ radians,*

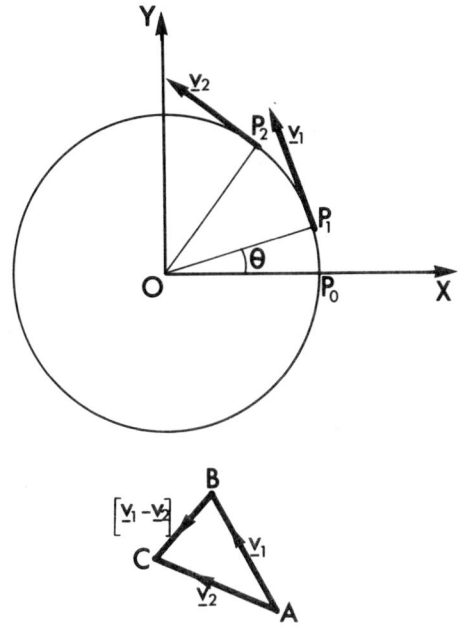

Fig. 1.3
Velocity and acceleration of a point moving in a circle.

$$\theta = \frac{s}{r} \qquad (1.5)$$

where r is the radius of the circle in metres. The speed of the point (v metre sec^{-1}) and the time taken (t sec) to move from P_0 to P_1 are related to s by

$$v = s/t \qquad (1.6)$$

---

* A *radian* is a useful way of measuring angles. One radian is the angle subtended at the centre of a circle by an arc whose length is equal to the radius of the circle. Since the circumference is 2π times the radius there are 2π radians in a complete revolution; that is, 2π radians = 360° or one radian = 57·3° approximately.

A quantity called the *angular velocity* ω is defined in an analogous way to eqn (1.6) by

$$\omega = \theta/t \tag{1.7}$$

Thus the angular velocity for the case of uniform circular motion is equal to the angle in radians swept out by the radius per second. Substituting eqns (1.5) and (1.6) into eqn (1.7) gives

$$\omega = (s/r)/(s/v) \tag{1.8}$$
$$= \frac{v}{r}$$

This is a useful relationship between angular velocity, speed and the radius. Note that since v and r are scalar quantities, so is the angular velocity ω. (It is possible to define angular velocity as a vector quantity but it is not necessary for us to do so in this book.)

As mentioned above, the velocity vector of the point is not constant because it is continually changing its direction. Therefore the point is accelerating, though this is not due to any change in speed because that remains constant. It is due to the continually changing direction of the velocity vector. We will calculate the magnitude of this acceleration and then deduce its direction. Suppose the point is at P_1 at time t_1 and P_2 at time t_2. Equation (1.4) gives us an expression for the acceleration vector

$$\mathbf{a} = \frac{(\mathbf{v}_2 - \mathbf{v}_1)}{t_2 - t_1}$$

when P_2 moves close to P_1 and $(t_2 - t_1)$ becomes small. In Fig. 1.3 the velocity vectors $\mathbf{v}_1$ and $\mathbf{v}_2$ have been re-drawn as a vector diagram. The difference $(\mathbf{v}_2 - \mathbf{v}_1)$ makes the third side of a triangle ABC. To show this is correct, think of AC as the resultant of vectors AB and BC; then, applying eqn (1.1), we get

$$\mathbf{v}_1 + (\mathbf{v}_2 - \mathbf{v}_1) = \mathbf{v}_2$$

Now because the velocity vectors are always tangential to the curve, $\mathbf{v}_1$ is perpendicular to OP_1 and $\mathbf{v}_2$ is perpendicular to OP_2. This means that angle BAC is equal to angle P_1OP_2. Also both triangle ABC and triangle P_1OP_2 are isosceles. Thus triangles BAC and P_1OP_2 are similar,

$$\therefore \frac{BC}{v} = \frac{s}{r} \tag{1.9}$$

Mechanics

(This is where the requirement that $(t_2 - t_1)$ be small is used. In eqn (1.9) the chord length $P_1 P_2$ has been taken as equal to the arc length and we have put v equal to the length of the sides AB and AC.) From eqn (1.9), $BC = \frac{vs}{r}$, so we can substitute $s = vt$ from eqn (1.6) to obtain $BC = \frac{v^2 t}{r}$, or since t is the time difference between P_1 and P_2,

$$BC = \frac{v^2 (t_2 - t_1)}{r} \tag{1.10}$$

Writing the acceleration in terms of magnitudes rather than vectors gives

$$a = \frac{BC}{(t_2 - t_1)} \tag{1.11}$$

Substituting for BC from eqn (1.10) yields

$$a = \frac{v^2}{r} = r\omega^2 \tag{1.12}$$

The direction of this acceleration is always towards the centre of the circle. This can be understood by imagining triangle ABC in the limit that $(t_2 - t_1)$ is very small. Angle BAC will be very small and BC will be approximately perpendicular to AB. Since $\mathbf{v}_1$ is directed along AB and this is tangential to the curve, BC must be along a radius directed towards the centre. This constant inward acceleration of an object moving in a circle is referred to as *centripetal acceleration*.

Example: Calculate the centripetal acceleration of a point on the earth's equator assuming that one complete revolution takes 24 hours and that the radius of the earth is $6·4 \times 10^6$ metres. In one revolution the earth turns through 360° or 2π radians. The angular velocity ω of a point on the equator is

$$\omega = \frac{2\pi}{24 \times 60 \times 60} \text{ radians sec}^{-1}$$

$$= 7·27 \times 10^{-5} \text{ radians sec}^{-1}$$

Using eqn (1.8) we can calculate the speed of a point on the equator as

$$v = 7·27 \times 10^{-5} \times 6·4 \times 10^6 \text{ metre sec}^{-1}$$

$$= 4·65 \times 10^2 \text{ metre sec}^{-1}$$

The centripetal acceleration is given by substituting in eqn (1.12),

$$a = \frac{(4\cdot 65 \times 10^2)^2}{6\cdot 4 \times 10^6} \text{ metre sec}^{-2}$$

$$= 3\cdot 4 \times 10^{-2} \text{ metre sec}^{-2}$$

This is small compared with the acceleration due to gravity which is about $9\cdot 8$ metre sec^{-2}.

Dynamics: Newton's laws of motion

In the last two sections we discussed methods of describing particles in motion. In this section we are going to discuss the reason why objects move at all.

This reason is simple; things start to move, or slow up, or change their direction of motion, when *forces* act. We all know intuitively what forces are because they are the basis of a great deal of our sensory experience. We kick a stone and it flies off in some direction — if it strikes a window the force applied to the glass is generally catastrophic in its effects. So the force of our contracting leg muscles eventually shatters the window into fragments via an intermediary, the stone, which was first accelerated and then decelerated. Clearly a force will cause acceleration, and conversely an oppositely-directed force (the resilience of the window) will cause deceleration. Only if no force is acting will velocity stay constant. These intuitive ideas were first explicitly stated by Isaac Newton (1642–1727) in three laws which bear his name.

Before we discuss Newton's laws in detail we should say a word about the coordinate system in which the various vectors involved (acceleration, velocity, force and others) are described. Consider, for instance, the force of gravity. As you sit reading this the pull of the earth probably feels much the same to you as it has done all your life. It has a certain magnitude and is directed vertically downwards. But under certain circumstances (travelling in a lift or elevator, flying in a plane that hits an 'air pocket') gravity feels very different — sometimes stronger, sometimes weaker. The extreme situation is that of astronauts circling the earth, for whom there is *no* apparent pull of gravity. The common feature of all these cases is that in each one the person involved is accelerating, either up or down; the astronauts of course have their centripetal

Mechanics

acceleration towards the centre of the earth, and in their case it is exactly equal to the acceleration of gravity. They are in effect (and it is so described) in 'free fall'. Since forces produce accelerations it is reasonable that, conversely, accelerations should produce 'forces'. Whether we should call such forces 'real' or 'spurious' is immaterial here, because they have very obvious effects. For instance, tie a stone on a string and spin it around your head. Since the stone is accelerating inwards (centripetal acceleration again) there must be some force acting causing it to do so. This force is felt as a tension in the string; cut the string, and both force and acceleration disappear and the stone carries on in a straight line. This particular force, connected with circular motion, is called *centrifugal* force and it is felt in this case as a tug on your hand.

Clearly, then, when we write down any relation between a force and the acceleration it produces we should ensure as far as possible that there are no other accelerations present. We saw above that the centripetal acceleration at the earth's surface, even on the equator, is weak compared with gravity, so any ordinary set of coordinate axes, fixed to the earth, will do quite well. But we should not, for instance, fix our axes to a circling spacecraft — the laws of physics look rather peculiar viewed by a freely-fall astronaut! (For a fuller discussion of this and allied topics, see the book by Lindsay and Margenau, p. 219.)

After these preliminaries, Newton's laws themselves perhaps seem almost self-evident, but in the seventeenth century they were startlingly novel. We shall now discuss each in turn.

Newton's first law of motion states that a body with no forces acting on it moves without acceleration. Thus a stone slid along a horizontal surface would continue to move with a constant velocity for ever were it not for the existence of frictional forces which slow it down. (Of course these frictional forces are always present, even for example on an ice surface. Thus the law was a very daring extrapolation from everyday experience.)

It is useful to rephrase Newton's first law in terms of a quantity called momentum (**P**). The momentum of a particle is defined as the product of its mass m kilograms and its velocity **v** metre sec^{-1}. Thus

$$\mathbf{P} = m\mathbf{v} \qquad (1.13)$$

Since velocity is a vector and mass a scalar quantity, momentum is a vector quantity. Newton's first law can now be re-phrased as — a body on which no force is acting moves with constant momentum. This follows because zero acceleration implies constant velocity and hence constant momentum.*

One of the reasons why momentum is a useful quantity is that it is conserved (i.e. remains constant) when particles interact. This is called the *law of conservation of momentum*. Suppose two particles collide. Call the momentum of the particles before the collision $\mathbf{P}_1$ and $\mathbf{P}_2$ and after the collision $\mathbf{P}'_1$ and $\mathbf{P}'_2$. Then it is experimentally observed that the sum of the momentum vectors before the collision is equal to the sum of the momentum vectors after the collision. Thus

$$\mathbf{P}_1 + \mathbf{P}_2 = \mathbf{P}'_1 + \mathbf{P}'_2 \tag{1.14}$$

This fact remains true no matter how many particles are involved in the interaction.

Equation (1.14) can be re-written as

$$\mathbf{P}_1 - \mathbf{P}'_1 = -(\mathbf{P}_2 - \mathbf{P}'_2) \tag{1.15}$$

The term on the left-hand side is the change in momentum of particle 1. Let this be denoted by $\Delta \mathbf{P}_1$. Similarly the term on the right hand side is minus the change in momentum of particle 2, $\Delta \mathbf{P}_2$.

$$\therefore \quad \Delta \mathbf{P}_1 = -\Delta \mathbf{P}_2 \tag{1.16}$$

In words, the momentum lost by particle 1 is exactly equal to that gained by particle 2.

Example: A 70 kg tennis player serves a 70 g tennis ball at 50 m s^{-1}. The momentum of the ball changes from zero to $70 \times 10^{-3} \times 50 = 3 \cdot 5$ kg m s^{-1}, and so the server gains an equal and opposite amount of momentum, recoiling from the direction of the serve with a velocity of $3 \cdot 5/70 = 1/20$ m s^{-1}.

The law of conservation of momentum is particularly important in explaining the locomotion of animals. The underlying principle

---

* The mass of a body is assumed to be constant in this treatment. In reality the mass of a body does increase with its velocity but this only becomes significant at velocities approaching the speed of light in a vacuum (roughly 3×10^8 metre sec^{-1}). This increase in mass with velocity ensures that no particle can travel with a velocity greater than that of light in vacuum and is a consequence of Einstein's theory of relativity. However, in the 'everyday world' velocities are not great enough for this variation of mass to be important and it will be ignored in this chapter. For an elementary introduction to relativity one cannot do better than read Einstein's own 'layman's guide' (p. 219).

Mechanics

of all animal motion is to move forward by pushing some part of the surroundings backwards. Fish swim by thrusting water back with their tails and fins; birds (and man-made rockets) soar in the air by giving a large amount of downward momentum to their environment. The birds give this momentum to the air directly by beating their wings, whereas rockets expel part of their own mass as burnt fuel, and are thus able to operate outside the atmosphere. For a more detailed discussion of the topic of animal locomotion, see the book by Gray (p. 219).

Returning to eqn (1.16), if the interaction of particles 1 and 2 takes place during a short time interval Δt (e.g. two billiard balls colliding) then dividing both sides of eqn (1.16) by Δt yields

$$\frac{\Delta \mathbf{P}_1}{\Delta t} = -\frac{\Delta \mathbf{P}_2}{\Delta t} \qquad (1.17)$$

We shall shortly find that this equation is useful in formulating Newton's third law.

Whilst on the topic of momentum it is worth mentioning that an analogous quantity can be defined for the case of circular motion. It is known as *angular momentum* and, like the linear momentum just discussed, is also a conserved quantity. The angular momentum of a mass m moving in a circle of radius r with velocity v is mvr, equal to $mr^2\omega$ where ω ($= v/r$) is the angular velocity. The fact that angular momentum is conserved is often obscured in nature by the ever-present frictional forces which tend to damp out rotational motion as soon as it starts; witness how quickly a stirred cup of coffee stops rotating. One case where friction is minimised is that of skaters or ballet dancers who can utilise the conservation law to increase their angular velocity almost at will during a spin. The trick is to hug one's arms or legs to one's sides, thus placing more mass at a smaller radius. Since $mr^2\omega$ is constant and mr^2 decreases, ω must increase to compensate. We shall make further use of angular momentum in Chapter 6, but for the present we return to linear momentum.

Newton's first law states what happens in the rather special case that no forces are acting, and tells us that no accelerations occur; clearly if forces are acting then accelerations will occur, and *Newton's second law* says simply that the acceleration so produced is proportional to the force applied. In fact Newton went further. Clearly a given force will accelerate a heavy body less than a light one, and Newton saw that the force could be written as the product of *mass* and acceleration, thus:

$$\mathbf{F} = m\mathbf{a} \qquad (1.18)$$

or, in words, 'force equals mass times acceleration'. This is one of the most fundamental equations in science and we shall often make use of it.* Notice that since mass is a scalar and acceleration a vector, force is also a vector quantity. We have already tacitly assumed this by talking about the directions in which forces act. If m is measured in kg and a in m s^{-2} then the unit of F will be kg m s^{-2}, which unit is called the newton (N).

Referring back to eqn (1.4) we see that the acceleration of a particle is simply its change in velocity, $\Delta \mathbf{v}$, divided by the small time interval Δt in which this occurs. Thus we can rewrite the second law as

$$\begin{aligned}\mathbf{F} &= m \frac{\Delta \mathbf{v}}{\Delta t} \\ &= \frac{\Delta \mathbf{P}}{\Delta t}\end{aligned} \qquad (1.18a)$$

since momentum $\mathbf{P} = m\mathbf{v}$. So another way of expressing the second law is 'force equals rate of change of momentum'. Also, if we are dealing with the interaction of two particles then eqn (1.17) applies and so

$$\mathbf{F}_1 = -\mathbf{F}_2 \qquad (1.19)$$

that is the force exerted by particle 1 on particle 2 is the same in magnitude (but opposite in direction) as that exerted by 2 on 1. This is *Newton's third law*, sometimes stated as 'action and reaction are equal and opposite'. The interaction need not be an actual collision, as with billiard balls. For instance a magnet will attract a piece of iron, or pieces of fluff may be electrically attracted to a haircomb, without any physical contact. In each case the magnet (or comb) is equally attracted towards the iron (or fluff), as eqn (1.19) implies.

This leads us to consider what types of force exist. At first sight there are an exceedingly large number but one of the

---

* We have avoided saying what mass *is*. Probably most readers are happy to accept that it is in some way related to amount of matter, but in fact its precise definition is an interesting metaphysical problem. For a clear discussion of the issues involved, see the book by Lindsay and Margenau.

Mechanics

successes of modern physics has been to reduce the total number of basically different kinds of force to only four. One of these is the *gravitational* force; the whole subject is (probably apocryphally) supposed to have begun when an apple fell on Newton's head.* We shall return to gravity shortly. Then there are two rather mysterious kinds of force which appear to operate only inside an atomic nucleus. We shall talk about them in Chapter 7; their consequences for our everyday world are negligible. Finally, by far the most important force is the *electromagnetic* force. Although electric and magnetic forces seem to differ in kind, we know now that they are only varieties of a single basic effect. Electrical forces hold atoms together and prevent the electrons and the nucleus from flying apart; more than this, all forces between molecules, such as van der Waal's effects and ionic bonding, are electromagnetic in character. Since these intermolecular forces are responsible for the tension of springs and muscles, the resilience of solids, and the friction between moving objects, we can truly say that the electromagnetic force determines the nature of the world we live in. That the effects are not even more apparent is simply because matter is, usually, electrically neutral, so that positive and negative charges balance exactly. Under these circumstances the major interaction between massive objects is due to gravitation, to which we now turn.

Newton's law of gravitation

This states that the force between two objects of masses M_1 and M_2, a distance R apart, is

$$F = \frac{GM_1 M_2}{R^2} \qquad (1.20)$$

where G is a constant having the value $6 \cdot 67 \times 10^{-11}$ N m² kg⁻². This is a universal law; all objects obey it, wherever they are, since so far as we know there is no gravitational analogy to an electrically neutral body. In particular the earth, of mass M_e say, attracts an object of mass M at a distance R from its centre according to eqn (1.20). If the object is at the surface of the earth, so that $R = R_e$, the earth's radius, we have

---

* Alexander Pope, a contemporary, wrote:
> Nature and Nature's laws lay hid in night:
> God said, *Let Newton be!* and all was light.

$$F = \left(\frac{GM_e}{R_e^2}\right)M \tag{1.21}$$

The quantity in brackets is called g, the acceleration due to gravity, and the equation can be rewritten as

$$F = mg \tag{1.22}$$

Due to irregularities in the earth's shape and density the value of g varies somewhat over its surface, its average being about 9·81 m s^{-2}. Any object allowed to drop towards the earth will, in the absence of air resistance, fall with this acceleration.

Example: A stone is thrown upwards with velocity 20 m s^{-1}. How high does it rise, and for how long does it stay in the air? We can tackle this as we did the example following eqn (1.4). First find the time before the velocity is zero. From eqn (1.4) this is $20/g \approx 2$ sec (since g is approximately 10 m s^{-1}). A similar argument holds as the stone falls back, so it stays in the air for a total of about 4 seconds. As for the height, we can substitute the average velocity (10 m s^{-1}) in eqn (1.3) to find the height = 10 m s^{-1} × 2 s = 20 m.

Now we can look again at the question of what keeps an orbiting satellite from falling to the earth. Such a satellite can remain aloft without any form of propulsion for years. Since the satellite will experience an attractive force towards the centre of the earth given by eqn (1.20) we know it will have an acceleration a towards the centre of the earth given by

$$a = \frac{GM_e}{r^2} \tag{1.23}$$

This follows since force equals mass times acceleration. Thus we know that the satellite will always be accelerating towards the centre of the earth with an acceleration a given by eqn (1.23). But it will be recalled that if a particle is moving in a circular orbit it has a centripetal acceleration towards the centre of the orbit, of magnitude given by eqn (1.12)

$$a = \frac{v^2}{r}$$

So by moving in a circular orbit with appropriate radius and velocity the satellite is able to be continually accelerating towards the centre of the earth with exactly the right value demanded by gravitational attraction. The condition that this be so is that the

Mechanics

right-hand sides of the last two equations be equal.

$$\therefore G\frac{M_e}{r^2} = \frac{v^2}{r}$$

$$\therefore v = \sqrt{\frac{GM_e}{r}} \qquad (1.24)$$

Not all satellite orbits are circular but the same principle applies to the actual orbits. A two hour period of rotation around the earth is not uncommon. To what radius of circular orbit does this correspond? The time per orbit T is given by

$$T = \frac{2\pi r}{v}$$

$$= 2\pi \sqrt{\frac{r^3}{GM_e}}$$

$$\therefore r = \left(GM_e \frac{T^2}{4\pi^2}\right)^{\frac{1}{3}}$$

Substituting $M_e = 5 \cdot 98 \times 10^{24}$ kilograms and $T = 2$ hours we get

$$\therefore r = \left(\frac{6 \cdot 67 \times 10^{-11} \times 5 \cdot 98 \times 10^{24} \times (2 \times 60 \times 60)^2}{4 \times (3 \cdot 14)^2}\right)^{\frac{1}{3}}$$

$$= 8 \times 10^6 \text{ metres}$$

Since the radius of the earth is about $6 \cdot 4 \times 10^6$ metres, the satellite's height would be roughly $1 \cdot 6 \times 10^6$ metres, or about 1000 miles.

Work and power

Whatever one's commonsense notion of 'work', in science the word has a very precise meaning. Work is said to be done when a force acting on an object displaces it. The work done is equal to the displacement times the component of the force along the displacement. Thus if a particle is constrained to move along a curve AB then the work done by a force **F** in moving it a short distance ΔS along the curve is equal to ΔS times the component of the force along ΔS. The distance ΔS is small enough for that region of the curve to be treated as a straight line. Force is a vector quantity and as we have discussed in the section on vectors any vector can be resolved into two component vectors. In our present example we choose one component to be parallel to ΔS and the other perpendicular to it. The magnitude of the component parallel to ΔS is $F \cos \theta$ where F is the magnitude of the force and θ is the angle between ΔS and **F** (see Fig. 1.4). Thus the work done, W, is given by

$$W = F \Delta S \cos \theta \qquad (1.25)$$

(The work done by the component of the force perpendicular to the displacement is zero. For example, there would be no work done by the force of gravity when a smooth stone is slid across horizontal ice; see Fig. 1.4.)

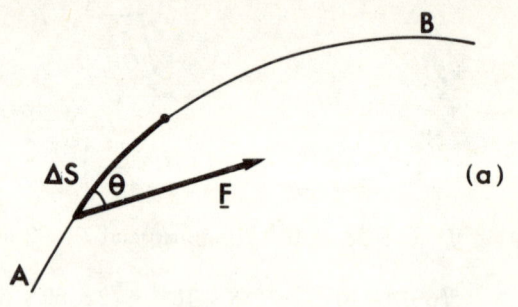

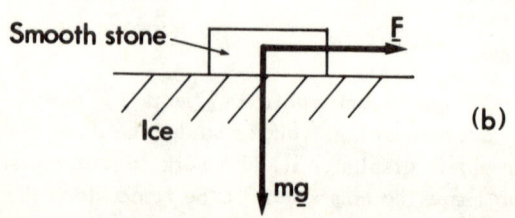

Fig. 1.4
(a) The work done in moving the particle a small distance ΔS along the curve is $F \Delta S \cos \theta$.
(b) The force mg due to gravity is perpendicular to the direction of motion and hence does no work.

Mechanics

In order to calculate the total work done in moving from *A* to *B* along the curve (Fig. 1.4) it is necessary to add up all the terms like the one given in eqn (1.25). Work, a scalar quantity, is measured in newton metres. One newton metre is called a joule (J).

If a machine or a living organism can do an amount of work ΔW joules in a short time Δt secs, then the rate of doing work per second P_W is

$$P_W = \frac{\Delta W}{\Delta t} \qquad (1.26)$$

This quantity P_W is called *power*. The unit of power is the *watt*. One watt is a rate of work of one joule per second. A multiple that is frequently used is the kilowatt, equal to 10^3 watts, and work is sometimes expressed as kilowatt-hours. Power, like work, is a scalar quantity.

Example: Suppose a 70 kg man climbs a flight of stairs 10 m high. The amount of work he does against the force of gravity is $mg \times$ height = $70 \times 9 \cdot 8 \times 10 \approx 7000$ J. If he runs up in 10 seconds his power output is 7000/10 = 700 watts (about one horse-power, an obsolete unit of power).

Often force is applied over an area rather than at a point. For example, a gas which is being compressed exerts a force on the compressing piston. In such a case it is usual to talk of the *pressure* exerted by the gas rather than the force. Pressure is defined as force per unit area. The unit of pressure is newton metre^{-2}.

It is interesting to consider how much work is done by a force compressing a gas which for simplicity we will imagine enclosed in a cylinder as shown in Fig. 2.1 (p. 38). Suppose that the piston has an area of A metre² and that it is depressed through a distance Δx metres thus reducing the volume by a small amount ΔV

$$\Delta V = A \Delta x \qquad (1.27)$$

The work done, *W*, is

$$W = F \Delta x$$

but

$$F = PA$$

by definition of pressure.

$$\therefore W = P A \Delta x$$

Using eqn (1.27) this becomes

$$W = P\Delta V \tag{1.28}$$

Thus the work done in decreasing the volume by a small amount ΔV (so small that the pressure can be taken as constant) is equal to the product of the pressure times the volume decrease.

Let us, by way of example, calculate the work done in exhaling air from one's lungs. The pressure difference we will assume to be constant and equal to some fraction, say one-tenth, of an atmosphere, that is about 10^4 N m^{-2}. The volume of air breathed out can be as much as 1 litre = 10^{-3} m³. Thus the work done is $10^4 \times 10^{-3}$ = 10 J. If one breathes out once every 4 seconds his rate of work is $10/4 = 2\frac{1}{2}$ W. By comparison the power needed to keep the human body at its normal temperature, and provided by the metabolism of foodstuffs, is about 100 W, so the extra effort involved in breathing is relatively negligible.

Energy

Energy may be thought of as the ability to do work. When we do work on an object we add to it an amount of energy which is equal to the work done. The units of energy and work are the same, namely the joule. In this section we will discuss two types of mechanical energy, *kinetic energy* which is energy of motion and *potential energy* which is energy of position. There are many other forms of energy as we shall see later, e.g. electrical, magnetic, chemical and nuclear energy, and all forms of energy are mutually inter-convertible. Sometimes, of course, the conversion of one form into another is not easy; building a nuclear reactor to produce electrical energy is not simple. But since most forms of energy are fairly easily converted into kinetic energy, and since its measurement is relatively easy also, we shall begin with this form.

Any moving object possesses kinetic energy; clearly work has to be done to stop it. If we wished we could simply state without proof the expression (eqn 1.33) which gives a body's kinetic energy in terms of its mass m and velocity v, but it is instructive to derive it using our earlier results. Suppose a force F is applied to the object, a stone, say, for a distance s. By Newton's law the stone has acceleration a given by

$$a = \frac{F}{m} \tag{1.29}$$

Mechanics

The stone starts from rest and finishes with speed v. The speed increases uniformly since the acceleration is constant, the average speed being half the difference between the final and initial speeds, $\frac{1}{2}(v - 0) = \frac{1}{2}v$. The time t for which the force is applied can be calculated by dividing the distance s by the average speed

$$t = \frac{2s}{v} \tag{1.30}$$

Since it has acceleration a during this time and starts from rest its final speed v is given by

$$v = at \tag{1.31}$$

Substituting eqn (1.30) into eqn (1.31) gives

$$v^2 = 2as \tag{1.32}$$

Replacing a in eqn (1.32) by its value in eqn (1.29) gives

$$Fs = \tfrac{1}{2}mv^2 \tag{1.33}$$

The term Fs on the left-hand side is equal to the work done on the stone. The term $\tfrac{1}{2}mv^2$ is therefore the kinetic energy acquired by the stone as a result of the work done on it. The stone will be able to do an amount of work equal to $\tfrac{1}{2}mv^2$ in being stopped. (This way of showing that the work done on the stone is equal to $\tfrac{1}{2}mv^2$ is rather specific. It relies upon the force acting on the stone being constant over a distance s. The result, however, is perfectly general.)

Example: How much work is required to stop a 1500 kilogram car which is travelling at 28 metre sec^{-1} (i.e. about 100 kilometres per hour)? The amount of work required is equal to the kinetic energy of the car. It will be done by the brakes and will turn into heat mostly in the brake linings. The kinetic energy

$$= \tfrac{1}{2} \times 1500 \times (28)^2 \text{ joules}$$
$$= 59 \times 10^4 \text{ joules}$$

The kinetic energy of any moving object, from a molecule to a racehorse, may be calculated using eqn (1.33).

Now let us turn to *potential energy*, which is really a blanket term covering all forms of energy that can be converted fairly directly into kinetic energy. For instance, consider a stone at rest above the ground. When the stone is released it will accelerate at 9·8 metre sec^{-2} until it strikes the ground. As it reaches the ground, because it has a velocity, it also has kinetic energy which could be used to do work. Clearly when the stone was stationary above the ground it had the capacity to do this work. The amount of work which it was capable of doing is called its potential energy.

The potential energy of a mass m at height h above the ground is easily calculated. Imagine raising it from the ground, vertically to a height h. The force pulling the mass towards the earth is a constant, mg, that is the mass times the acceleration due to gravity. Therefore the work done, W, is given by

$$W = (mg) \times h$$

$$\therefore \text{potential energy} = mgh \tag{1.34}$$

When the mass is dropped, h becomes zero on reaching the ground so all the potential energy has been converted into kinetic energy

$$\therefore mgh = \tfrac{1}{2} mv^2 \tag{1.35}$$

Therefore its speed on reaching the ground is

$$v = \sqrt{2gh} \tag{1.36}$$

The fact that the speed is independent of the mass is not observed in reality. A feather falls more slowly than a stone, due to air resistance which we have ignored in the above treatment. A striking demonstration of this is the simple experiment of dropping a feather and a coin in an evacuated bell-jar, and seeing them hit the bottom together.

Equation (1.35) could have been derived in a slightly more general way. It follows directly from the fact that the total energy of the mass, that is the sum of kinetic and potential energy, is constant. The mass starts at rest and finishes with zero potential energy so eqn (1.35) follows immediately. We shall return to this shortly.

We have been discussing gravitational potential energy but there are numerous other examples. A compressed spring has potential energy and so has a piece of iron pulled a short distance away from a magnet. In both cases it is clear that if the system is released motion will occur as the potential energy is converted into kinetic energy.

We should perhaps state here that for some forms of energy units other than the joule are still quite commonly used. The *electron-volt*, for instance, is used to measure the minute quantities of

Mechanics 23

energy in single atoms (see Chapter 6). Chemists, too, are prone to measure the energy of chemical reactions in *kilocalories per mole* of reactants; a calorie is a unit equal to 4·184 J so a kilocalorie is 4184 J.

As an example consider a tree from whose leaves water is being evaporated. This water is forced up* from ground level, thus acquiring potential energy which is ultimately derived from sunlight via the breakdown of photosynthesised carbohydrates. Suppose 1 kg of water to be raised to a height of 50 m. The potential energy acquired is $1 \times 9 \cdot 8 \times 50 \approx 500$ J, corresponding to about 1/8 of a kilocalorie. The breakdown of carbohydrates commonly yields chemical energies of several hundred kilocalories per mole so only a very small fraction of a mole (typically less than a gram) has to be broken down to pump 1 kg of transpired water.

Let us consider now another example, this time concerned with electrical energy. After stating some facts about the force exerted by an electric charge, we shall show that an electron can move in a circular orbit around a nucleus, just as a planet circles the sun, and we shall calculate the kinetic energy of the electron in this orbit. The expression we derive will be useful in Chapter 6, and if desired this section could be omitted on first reading.

An electric charge of q coulombs is surrounded by an *electric field* whose strength, E volts per metre, at a distance r from the charge is given by

$$E = \frac{q}{4\pi\epsilon_0 r^2}$$

where ϵ_0 is a constant (*permittivity of a vacuum*) having the value $8 \cdot 85 \times 10^{-12}$ farad metre^{-1}. The consequence of this electric field is that another charge of q' coulombs placed at a distance r metres experiences a force of F newtons given by

$$F = Eq' = \frac{qq'}{4\pi\epsilon_0 r^2}$$

This force is directed along the straight line joining the centres of the charges. It is a repulsive force for like charges (+ + or − −) and attractive for unlike charges (+ −). Now consider two like charges, one of them fixed in position. In order to bring the other charge from

---

* The problem of *how* water finds its way to the tops of tall trees is by no means solved yet; see Chapter 2 (p. 57) and the references cited there.

infinity to a distance r metres away it is necessary to do work. This work is equal to the potential energy of the moveable charge. It can be regained if the moveable charge is released. It would then be accelerated away (towards infinity) and would finish up with a kinetic energy equal to its initial potential energy.

Consider an electron having charge $-q$ coulombs moving in a circular orbit of radius r metres around a nucleus of charge $+q'$ coulombs. Let m be the mass of the electron. The electron will be accelerated towards the nucleus because of the force exerted on it by the electric field of the latter. The force F is given by the above equation, so the acceleration a of the electron is

$$a = \frac{F}{m}$$

$$= \frac{qq'}{m \cdot 4\pi \epsilon_0 r^2}$$

Now the electron can have this acceleration towards the nucleus and yet never get any closer by moving in a circular orbit of appropriate radius with a suitable speed. The radius r and speed v must be such that the centripetal acceleration $\frac{v^2}{r}$ is equal to the acceleration caused by the electrical attraction.

$$\frac{v^2}{r} = \frac{qq'}{4m\pi \epsilon_0 r^2}$$

$$\therefore \tfrac{1}{2}mv^2 = \frac{qq'}{8\pi \epsilon_0 r}$$

This gives a connection between the kinetic energy of the electron in its orbit and the radius of the orbit. It tells us that an orbit of any radius is possible provided the kinetic energy of the electron satisfies the above equation. Note that the technique we have used is exactly the same as that used in discussing satellites moving under a gravitational force.

Conservation of energy

We have mentioned that momentum is conserved. It is also found experimentally that energy is conserved. The *law of conservation of energy* states that the total amount of energy in a system isolated from its surrounds always remains constant although energy

Mechanics

transformations from one form to another may occur within the system. For example, the sum of the kinetic and potential energies of a mass falling due to gravity is a constant, a fact that can be used to solve innumerable problems of everyday life. Equating the initial potential energy, mgh, to the final kinetic energy $\frac{1}{2}mv^2$ (since the final potential energy may be taken as zero if the height is zero) we find $v = \sqrt{2gh}$, an expression we have already derived (eqn 1.36). This velocity is independent, not only of the mass but also of the path taken on falling — routes vertically downwards and around a curving helter-skelter are equivalent so far as final velocity is concerned (assuming, of course, that one can ignore friction). Suppose we drop a stone down a 100 m deep well. Its velocity on striking the water is $\sqrt{2 \times 9 \cdot 8 \times 100} = 44 \text{ m s}^{-1}$, and so its average velocity was 22 m s^{-1}. It thus took $100/22 = 4 \cdot 55$ s to fall; and since the velocity of sound is 333 m s^{-1} it will take another $0 \cdot 3$ s for the sound to reach us. Thus we hear the splash $4 \cdot 85$ seconds after letting go of the stone.

Further examples of the use of this conservation law are given in the exercises. It is a law which scientists would be very loath to give up; indeed it has been said that if a process were observed in which energy seemed to disappear, we would invent a new form of energy to account for it. (This attitude is not as irrelevant as it may seem. The *neutrino*, an elementary particle studied by nuclear physicists, was discovered only after it had seemed for some time as though energy conservation had broken down — the discovery of the neutrino saved the day. See Chapter 7 for a further discussion of this point.) Even nuclear energy, of which seemingly inexhaustible amounts are available, obeys this law. Einstein, in his theory of relativity, showed that a mass M could give an amount of energy E equal to Mc^2 where c is the velocity of light. This incidentally is the source of the sun's energy. Every second the mass of the sun decreases by about one million tonnes as it is transformed to radiant energy (happily this constitutes only about one part in 10^{21} of the sun's total mass).

As one further example of the use of the laws of conservation of energy and momentum we will consider the decay of a heavy nucleus by alpha particle emission. Alpha particle decay is discussed in Chapter 7. For our purposes here it is only necessary to know that a heavy nucleus at rest can split up into an alpha particle and a lighter (daughter) nucleus with a release of energy. The alpha particle and the daughter nucleus share the released energy between them as kinetic energy, thus conserving energy. The momentum of the initial heavy nucleus was zero so it is necessary for the daughter nucleus and the alpha particle to fly apart with equal and opposite momenta thus conserving momentum.

Let the energy released in the decay be E and the masses of the alpha particle and daughter nucleus be m_α and m_N respectively. Call the velocities of the alpha particle and the daughter nucleus v_α and v_N respectively. The conservation of energy requires that

$$E = \tfrac{1}{2} m_\alpha v_\alpha^2 + \tfrac{1}{2} m_N v_N^2$$

The conservation of momentum requires that

$$m_\alpha v_\alpha = m_N v_N$$

Eliminating v_N from the above two equations gives

$$E = \tfrac{1}{2} m_\alpha v_\alpha^2 + \tfrac{1}{2} \frac{m_\alpha^2}{m_N} v_\alpha^2$$

$$\therefore \tfrac{1}{2} m_\alpha v_\alpha^2 = \frac{E}{1 + \dfrac{m_\alpha}{m_N}} \tag{1.37}$$

Equation (1.37) is an expression for the kinetic energy of the alpha particle. There are two things to notice about it. First, the kinetic energy of the alpha particle is fixed. It is exactly determined by the energy release and the masses of the final particles. This means of course that the kinetic energy of the daughter nucleus is also fixed. So if a nuclear decay is observed where the kinetic energy of one of the particles is not unique (but for example varies from zero up to some maximum) then the decay can not be 'two-body' decay. There must be three or more objects (colloquially called 'bodies') produced by the decay. The second thing which we can learn about the decay is that since $\dfrac{m_\alpha}{m_N}$ is small the term $\left[1 + \left(\dfrac{m_\alpha}{m_N}\right)\right]$ is only slightly greater than unity. Thus we see from eqn (1.37) that the lighter particle in the decay (the alpha particle) gets most of the energy. So by simply applying the laws of conservation of energy and momentum to the decay we have learnt some things which were not immediately apparent.

Vibrations and waves

Of the many different types of motion encountered in nature one of the most important is *vibrational* motion. Trees sway to-and-fro in a breeze, winds cause ripples on water surfaces, and even the cilia of minute organisms beat, as do our hearts, in an oscillatory manner. Evidently there is a great deal in common between vibrational motion and *wave* motion. In this section we shall discuss the two in sequence, providing the background to a more detailed treatment in Chapter 5.

There are many types of vibrational motion, of varying degrees of complexity. Fortunately one of the simplest, and the easiest to treat mathematically, is also the commonest. It is called *simple harmonic motion* (nearly always abbreviated to SHM) and is exemplified by the way a pendulum swings. We shall define SHM, and introduce further examples from nature, in a moment, but we begin with a simple physical example to show its main features.

Consider a particle of mass m which is attached to a wall by a spring. This mass is able to slide on a horizontal frictionless surface and is held so that the spring extension is x_0 (Fig. 1.5a). It is necessary to do work in order to stretch the spring, thus giving it potential energy. When the mass is released it will be accelerated towards the wall by the force exerted on it by the stretched spring. Thus, due to the conservation of energy, the mass gains kinetic energy and the spring loses an identical amount of potential energy. The situation at this stage is portrayed schematically in Fig. 1.5(b), where the mass is shown as having speed v and the spring an extension less than x_0.

Eventually the mass reaches the position corresponding to zero spring extension. The speed of the mass has increased to its greatest value v_{max}. At this stage the potential energy is zero because the spring extension is zero. All the potential energy of the spring has been converted to kinetic energy, which now has its maximum value (Fig. 1.5c). Clearly since the mass has kinetic energy it overshoots this position and continues moving towards the wall, compressing the spring. During this process the spring exerts a retarding force on the mass, which slows it down. Thus some of the kinetic energy of the mass is converted back into potential energy of the spring (Fig. 1.5d). Eventually this process is complete and the mass stops (Fig. 1.5e). At that stage all the kinetic energy of the mass has been converted into potential energy of the spring.

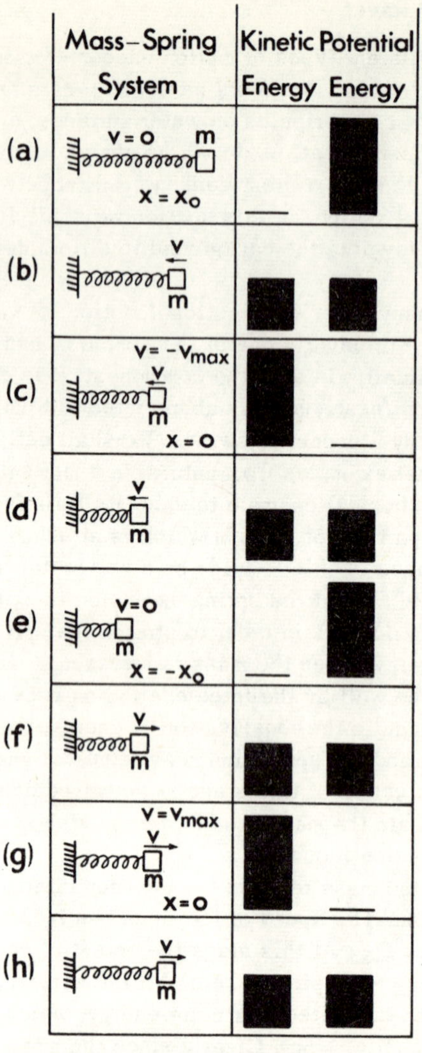

Mass–Spring System	Kinetic Energy	Potential Energy
(a) $v=0$, m, $x = x_0$	—	■
(b) v ←, m	■	■
(c) $v = -v_{max}$, m, $x = 0$	■	—
(d) v ←, m	■	■
(e) $v = 0$, m, $x = -x_0$	—	■
(f) v →, m	■	■
(g) $v = v_{max}$, m, $x = 0$	■	—
(h) v →, m	■	■

Fig. 1.5
Simple harmonic motion of a mass-spring system. Time increases by equal increments from (a) to (h). After (h) the motion repeats from (a) again.

The spring is now compressed by an amount x_0. The mass then starts to accelerate away from the wall due to the force of the compressed spring (Fig. 1.5f). It again overshoots the equilibrium

Mechanics

position corresponding to zero spring extension, and in fact continues to vibrate backwards and forwards.

The characteristics of the vibration are that the displacement of the mass varies from x_0 to $-x_0$. The acceleration varies in a similar manner, having its maximum values when the displacement is a maximum. A most important property of the motion is the interchange between the two types of energy, kinetic and potential, and this is characteristic of all types of vibration.

What requirements have to be satisfied in order that the vibration outlined above is of the specially simple type we call SHM? It turns out that the only requirement is that the force exerted by the spring is of the form

$$\mathbf{F} = -k\mathbf{x} \qquad (1.38)$$

where k is a constant, $\mathbf{x}$ is the (vector) spring extension and $\mathbf{F}$ is the force. There are two facets to eqn (1.38). Firstly it states that the force exerted by the spring must be in such a direction as to decrease the spring extension. Secondly the force is proportional to the extension. This is known as *Hooke's law* and is satisfied by most springs provided their extension is not too large. In fact eqn (1.38) describes a wide range of forces in nature which is why SHM is so important. For instance, the stretched strings of stringed instruments, the parchment sheets of drums and the air columns of wind instruments all obey Hooke's law, so the theory of musical instruments is just the theory of SHM.

We have outlined the general features of simple harmonic vibrations. Some knowledge of the specific details will be useful in Chapter 5, so we give their derivations here. A particular example which represents a SHM will be described because it leads in a simple way to the mathematical description of SHM. Let us consider a point P moving with constant angular velocity ω around a circle in an anti-clockwise direction (Fig. 1.6). Its position at any instant is fixed by the vector OP emanating from the origin. The X component of the vector OP moves with a SHM. When the point P is on the X axis, the X component of OP is a maximum, corresponding to Fig. 1.5(a) or Fig. 1.5(e). When P is on the Y axis, the X component of OP is zero, corresponding to Fig. 1.5(c) or Fig. 1.5(g). More detailed consideration shows the correspondence between the vibration of the mass-spring system and the X component of OP holds in other respects as well. In other words, the motion of the point P' along the X axis is simple harmonic, if P goes

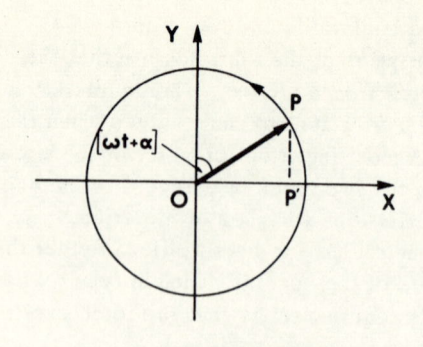

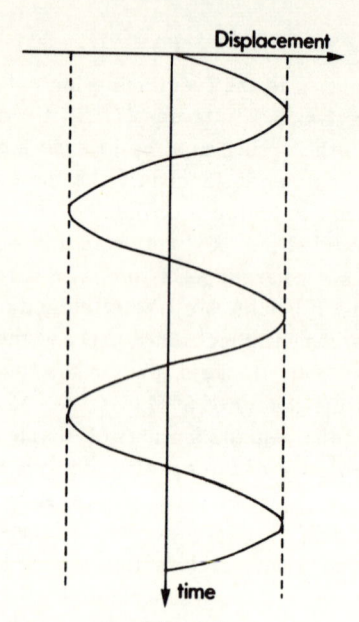

Fig. 1.6
The displacement along the X axis of a point moving in a circle represents a simple harmonic motion.

round the circle at a constant speed. A little trigonometry then gives the equation relating the distance x to the time.

Suppose the vector OP was at an angle α to the Y axis at time $t = 0$. Then at a later time t, OP will be at an angle $(\omega t + \alpha)$. From the geometry of Fig. 1.6 it can be seen that the X component of OP is given by

Mechanics

$$x = x_0 \sin(\omega t + \alpha) \qquad (1.39)$$

where x_0 is the length of OP, that is the radius of the circle. This is the general expression for SHM as a function of time. The quantity x_0 is termed the *amplitude* of the vibration.

The time for one complete vibration, that is the time for P to travel once around the circle, is called the *period* T of the vibration.

$$T = \frac{2\pi}{\omega} \qquad (1.40)$$

For the mass-spring system one can prove that

$$T = \frac{1}{2\pi}\sqrt{\frac{k}{m}} \qquad (1.41)$$

where k is the *force constant* of eqn (1.38). By using eqns (1.40) and (1.41) we can write down an expression for ω in terms of k and m. Equation (1.39) with this value of ω then describes the time-variation of the spring extension x.

It is interesting to note that the period of the vibration is independent of the amplitude of the oscillations. This feature is true also of the pendulum, a fact which has been utilised by clockmakers over the years (and which is said to have been noted first by Galileo, watching the swinging of a bronze lamp in the cathedral at Pisa).

The expression $(\omega t + \alpha)$ is called the *phase* of the vibration, α being the initial phase. Clearly, when the phase increases by 2π radians ($= 360°$) the sine function repeats itself, since one complete revolution has occurred.

One can deduce a great deal about a particular SHM simply by knowing something about the force constant k. Consider, for instance, a stretched wire which is plucked or struck to make it vibrate. The restoring force k is clearly the (small) component of the tensional force, F, which is pulling the wire back to its equilibrium position. Forgetting for the moment any numerical constants we can substitute this in eqn (1.41) and find that the period of the vibration is proportional to $\sqrt{F/m}$ where m is the mass of 1 metre of wire. This shows the manner in which the frequency ('pitch') of stringed instruments can be altered by adjustment of the string 'weight' m and tension F.

Another example can easily be tested by the reader. Hold a piece of glass tubing, open at both ends, half in and half out of a bowl of water. Stir the water to make it rise and fall in the tube.

The motion is SHM, the restoring force being due to the water 'finding its own level', and being equal to the gravitational force on the water. Thus $k = mg$ and eqn (1.41) simplifies to

$T = \frac{1}{2\pi}\sqrt{mg/m} = \frac{1}{2\pi}\sqrt{g}$. Substituting $g = 9 \cdot 8$ m s^{-2} we find that

the period T is almost exactly $\frac{1}{2}$ second. Try it and see; the tube size is immaterial.

In many vibrations only single particles or small groups of particles* are involved. Often, however, it happens that one particle transfers some of its energy to an adjacent one, which then passes it on to the next and so on. Thus a *wave* is set up in the material. Note that we specify that waves transfer energy from one place to another without transferring any matter. The particles, though vibrating, stay close to their average positions. For example, a sound wave travelling through air consists of a series of compressions and rarefactions which move through the air with the velocity of sound. Individual particles in the air oscillate only backwards and forwards. Similarly, the bobbing of a cork on the sea shows that the water does not move bodily (except at the shore, where the abrupt boundary upsets the conclusions drawn from our simple assumptions).

There are two types of waves, *longitudinal waves* and *transverse waves*. The distinction arises from the direction of motion of individual particles in the medium relative to the direction of the wave. In transverse wave motion, the individual particles move at right angles to the wave direction, as in water waves on the sea. In longitudinal wave motion, particle motion is parallel to the wave direction, as occurs in sound waves.

The connection between vibrational and wave motion can be seen in the following way. Focus attention on an individual particle executing a simple harmonic motion as it vibrates back and forth (Fig. 1.6). Next, rather than looking at one particle only as time proceeds, concentrate attention on one instant of time and look at all the particles. This could be thought of as taking a photograph of the wave motion using a very short exposure time. The result for a wave on a stretched string is shown schematically in Fig. 1.7.

---

* Although we have discussed vibrations from the point of view of particle motion, any physical quantity can obviously oscillate about some value or other, and these constitute perfectly good vibrations. Thus temperature goes up and down diurnally on the earth's surface, and this leads to waves of temperature moving into the earth from the surface. We shall return to non-material vibrations in Chapter 5.

Mechanics

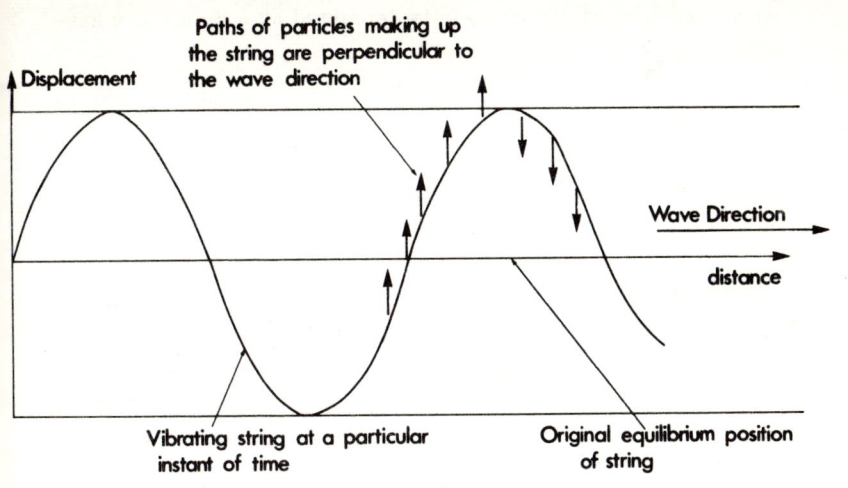

Fig. 1.7
Vibration of a string, showing relationship to wave motion.

The displacement of the string from its equilibrium position varies in a simple harmonic manner with distance down the string. The distance between the wave crests, the repeat distance in Fig. 1.7, is called the *wavelength* of the wave. As time proceeds the wave pattern moves in the direction of propagation of the wave. (This would be clear if a movie film of the string were taken.) This has the consequence that the displacement at any one point varies in a simple harmonic way as shown in the lower part if Fig. 1.6. The velocity of the wave v is related to the wavelength by

$$v = \lambda \nu \qquad (1.42)$$

where ν is the *frequency*, $\nu = 1/T$. Equation (1.42) is easily proved. In one second the wave travels a distance v. At the same time a total of $1/T = \nu$ complete wavelengths, each of length λ, pass by. Thus ν times λ must be equal to the total distance covered, v.

Phase, it will be recalled, is the argument of the sine function (that is, the angle whose sine is taken) which describes the SHM. It thus determines the distance from the origin of the wave, and hence the amplitude of the wave at that point. Thus if a circulars wave were spreading out, say from a stone dropped in a pond, then everywhere on the same wavecrest or trough has the same phase, since each such point is at the same distance from the centre.

At a fixed instant of time, points separated by a distance x have a *phase difference* of δ where

$$\delta = \frac{2\pi x}{\lambda} \tag{1.43}$$

To check that this is correct notice that when $x = \lambda$, the repeat distance of the wave, then $\delta = 2\pi$ and, as has been pointed out, adding 2π to the argument of the sine function leaves it unchanged.

Problem-solving in mechanics

Most problems in simple mechanics require the use of several of the concepts and laws we have discussed. In this section we give just two examples of how the equations are combined to get the final answer, as illustrations of the techniques commonly employed.

Example 1: Consider the question, if a man jumps at 30° to the horizontal with a velocity of 6 m s^{-1}, how far does he jump? Since no forces are acting horizontally (except air resistance which we ignore) the answer is, the distance that the horizontal component of his velocity will take him whilst he is in the air. We must first, therefore, calculate how long he stays off the ground. Only the vertical velocity component enters now, since gravity acts vertically. His initial vertical velocity is $6 \sin 30° = 3$ m s^{-1}, and his deceleration is g. Hence the time spent rising is (from eqn 1.4) $3/g$ seconds, and a similar time is spent falling. So the total time in the air is $6/g$ seconds leading to a horizontal distance travelled of (horizontal component) × (time) = $6 \cos 30° \times 6/g \approx 3\cdot 2$ metres.

Example 2: This example also is to do with athletics. A shot putter hurls his 7 kg weight by leaping across a circle and then suddenly throwing the weight from his outstretched arm, transferring all his momentum to it and thus bringing himself to rest. Suppose an 82 kg athlete putts the shot 20 m by throwing it at 45° to the horizontal. Make a rough estimate of his power output as he straightens his arm and releases the shot.

Consider what is required. Power is the rate of doing work, so we need the total energy expended by the arm muscles, and the time taken. The former quantity is the difference between the kinetic energies before and after throwing the shot, and the latter can be estimated if we know the velocity of our athlete's hand at the moment of release. The problem therefore reduces to finding the

Mechanics

velocities of the athlete and the shot just before and just after the shot is thrown, and this is easily done by applying momentum conservation.

The procedure is therefore as follows. First find the velocity with which the shot is thrown, that is work Example 1 'backwards' for a throwing angle of 45° rather than 30°. This gives a velocity of $\sqrt{20g} \approx 14\,\text{m s}^{-1}$. Thus, by the law of conservation of momentum, the athlete and shot together had a velocity v just before throwing given by the equation

$$(82 + 7) \times v = 7 \times \sqrt{20g}$$

or $v \approx 1\cdot 1\,\text{m s}^{-1}$. So the total kinetic energy of the athlete and shot before the throw was $\frac{1}{2}(82+7) \times (1\cdot 1)^2 \approx 54\,\text{J}$, whilst after the throw the total kinetic energy is $\frac{1}{2}(7) \times (20g) \approx 686\,\text{J}$ (since the athlete himself, we have assumed, is at rest). The energy expended by the arm muscles is the difference of these two quantities, or about 632 J. Also, since his hand has a final velocity, just as the shot is released, of $14\,\text{m s}^{-1}$, his arm, being about 70 cm long must take of the order of 1/10th of a second to straighten (assuming a constant acceleration). Consequently a rough estimate of the power output during this short time would be 632/(1/10), that is 6320 W.

2

THE MOLECULAR NATURE OF MATTER

In this chapter we show how the notion of matter as composed of many identical molecular particles can lead, by application of the laws of mechanics, to an understanding of such processes as diffusion, osmosis and evaporation. We also briefly mention the molecular basis of chemical reactions.

The idea that all matter consists of molecules in continuous motion, constantly colliding and exchanging energy, has had philosophical appeal for centuries but was first put on a sound basis in the early nineteenth century by the experiments of Rumford and Joule who showed that heat could best be considered simply as a measure of this motion. One of the early successes of molecular theory was its use by Maxwell to explain the bulk properties of gases, such as the relation between pressure, volume and temperature. This topic is known as *kinetic theory* and we shall start with a brief review of it.

Kinetic theory of gases

Gases exert a pressure on the walls of their containers because of the constant molecular bombardment. Our eardrums are bombarded by air molecules, though our ears are, fortunately, just insufficiently sensitive to detect this continual patter. Solids exert no similar pressure, because in these materials the forces between molecules are strong enough to hold them in more or less fixed positions and prevent them from colliding with the container walls. We shall use the laws of mechanics to derive an expression for the pressure exerted by a gas, assuming that the molecules are very small, hard spheres which bounce elastically (without any loss of energy) off each other and the container walls.

Imagine a cylinder, fitted with a frictionless piston and containing N molecules of a gas, each of mass m, in a volume V (Fig. 2.1). The molecules collide with the piston and exert a force on it. The piston will only remain at rest if an equal and opposite force F is exerted, as shown. Let us apply Newton's law in the form of eqn (1.18a), stating that the force exerted by the molecules on the piston is equal to their total rate of change of momentum in the direction concerned, that is along the X axis in Fig. 2.1. But we know that this molecular force is also equal to the pressure, P, multiplied by the area A of the piston. Thus F = rate of change of momentum = PA. So to find P we need to calculate the total rate of change of momentum of the molecules, that is the momentum change in each collision multiplied by the number of collisions that occur per second. The calculation is very simple if we make the assumption that the molecular velocities, instead of being randomly distributed and changing after every collision, are all the same and equal to the *average molecular velocity*, $\bar{v}$. (The bar over the top is a conventional way of denoting an average value.) Although this assumption is not strictly correct it turns out that it gives almost the same answer as is obtained from a more rigorous treatment.

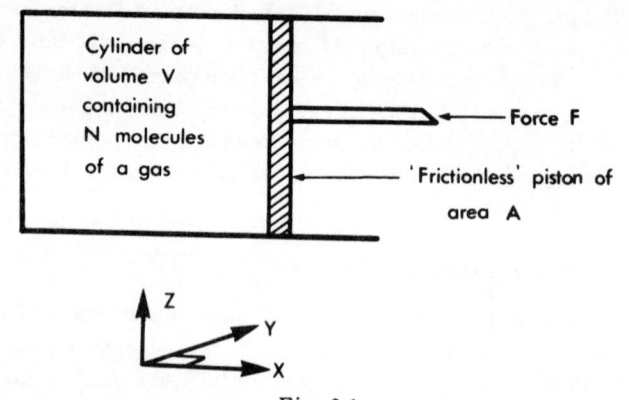

Fig. 2.1
A volume V of gas is contained in a cylinder by a 'frictionless' piston.

With this approximation, then, we can resolve $\bar{v}$ into components along the X, Y and Z axes, and call these $\bar{v}_x$, $\bar{v}_y$ and $\bar{v}_z$. Obviously during a collision with the piston only the X component of the velocity changes; in fact it reverses its direction, and so the momentum change in each collision is $m\bar{v}_x - (-m\bar{v}_x) = 2m\bar{v}_x$. Also, in one second all the molecules lying within a distance of $\bar{v}_x$ from the piston

The Molecular Nature of Matter

will strike it, as long as they are moving towards the piston rather than away from it, and on average one half of the molecules will be doing so. Thus, the number striking per second is one-half of the number of molecules per unit volume times the volume of a region a distance $\bar{v}_x$ away from the piston of area A, that is $\frac{1}{2}\left(\frac{N}{V}\right)A\bar{v}_x$.

Hence the rate of change of momentum is $(2m\bar{v}_x)\left(\frac{NA\bar{v}_x}{2V}\right) = \frac{mNA\bar{v}_x^2}{V}$, and so the rate of change of momentum per unit area is

$$P = \frac{F}{A} = \frac{mN\bar{v}_x^2}{V}$$

Now on average the components of velocity in the X, Y and Z directions will be equal, that is $\bar{v}_x = \bar{v}_y = \bar{v}_z$, and also by Pythagoras' theorem $\bar{v}^2 = \bar{v}_x^2 + \bar{v}_y^2 + \bar{v}_z^2 = 3\bar{v}_x^2$, say. Thus we can replace $\bar{v}_x^2$ by $\frac{\bar{v}^2}{3}$ to obtain

$$P = \frac{mN\bar{v}^2}{3V}$$

or $\qquad PV = \frac{1}{3}mN\bar{v}^2 = \frac{2}{3}N(\frac{1}{2}m\bar{v}^2) \qquad (2.1)$

This is the fundamental relation between the pressure and volume of a gas, and the average kinetic energy of molecular motion.

As was pointed out above, eqn (2.1) is not exactly correct. If we were to do the calculation properly, assuming a certain distribution of molecular velocities, and average over all the different kinetic energies at the end, we would obtain on the right-hand side the average of the velocity squared, $\overline{v^2}$, rather than the square of the average velocity, $\bar{v}^2$. It is clear that these two quantities may differ; if, for instance, the velocity varied sinusoidally with time its average would be zero (as often positive as negative) whereas its square has a non-zero average (always positive). For the random collisions with which we are concerned here it can be shown that $\bar{v} = \sqrt{\frac{3\pi}{8}}\sqrt{\overline{v^2}} = 1\cdot 18\sqrt{\overline{v^2}}$.

Thus an error of some 18% is involved in taking the two to be equal. The correct expression, then is

$$PV = \frac{2}{3}N\left(\frac{1}{2}m\overline{v^2}\right) \qquad (2.1a)$$

Example: To find the average velocity of hydrogen molecules at STP we look up the molecular mass and atmospheric pressure in suitable units and substitute in eqn (2.1a), using also the fact that one mole occupies 22·4 litres.

$$\text{Thus } \overline{v^2} = \frac{3PV_0}{mN_0} \simeq \frac{3(10^5 \text{Nm}^{-2})(0\cdot 0224 \text{m}^3)}{(3\cdot 2 \times 10^{-27}\text{kg})(6 \times 10^{23})}$$

$= 3\cdot 5 \times 10^6 \text{m}^2 \text{ sec}^{-2}$. Hence $\sqrt{\overline{v^2}} \approx 1\cdot 8 \text{ km sec}^{-1}$ and so $\bar{v} = 1\cdot 18 \sqrt{\overline{v^2}} \simeq 2\cdot 2 \text{ km sec}^{-1}$.

Having related the pressure and volume we now need to include the temperature to get a complete description of the properties of a gas. We define *temperature* as a quantity directly proportional to the average kinetic energy, $\frac{1}{2}m\overline{v^2}$, of the gas molecules. This identification is at the heart of kinetic theory and indeed accounts for the name. The concept was quite novel less than 200 years ago, Lavoisier for instance considering that rise in temperature was associated with the flow of a 'caloric' fluid into a body. It now seems very natural to link temperature with molecular motion, and to consider the flow of heat into a body as representing simply an increase in the molecular kinetic energies. Thus we write the average kinetic energy as*

$$\tfrac{1}{2} m\overline{v^2} = \tfrac{3}{2} kT \qquad (2.2)$$

where the constant of proportionality is $\frac{3}{2}k$ and k is known as *Boltzmann's constant*. Its numerical value is $1\cdot 38 \times 10^{-23}$ joules per °K and $\frac{3k}{2}$ represents the amount of energy which, for a single molecule, is equivalent to a one degree temperature change. The factor $\frac{3}{2}$ appears for numerical convenience; substitution of eqn (2.2) in eqn (2.1a) yields

$$PV = \tfrac{2}{3} N (\tfrac{1}{2} m\overline{v^2}) = NkT \qquad (2.3)$$

If we consider one mole of gas then

$$PV = N_0 kT = RT \qquad (2.4)$$

---

* Equation (2.2) defines not only the concept of temperature but also the *zero* of the temperature. This is the point at which the average kinetic energy has been reduced to zero, and on the Centigrade scale this occurs at −273·16°C. Temperature measured from this zero value is termed *absolute* temperature, and if the Centigrade scale is used the divisions are known as *degrees kelvin* (°K). Thus the freezing point of water is 0°C or 273·16°K.

The Molecular Nature of Matter

where N_0 is Avogadro's number and R, the *universal gas constant*, represents that amount of energy in one mole corresponding to a one degree temperature rise. R is simply $N_0 k$ and its numerical value is 8·31 joules per °K per mole, that is about 1·99 calories per °K per mole.

Equation (2.4), commonly called the *gas law*, embraces Boyle's law and Charles' law. Another law of great practical significance is *Dalton's law of partial pressures*, which may be proved very simply. Consider two gases, A and B, in a box of volume V. Let P_A, P_B, P_{AB} be the pressures exerted respectively by N_A molecules of gas A alone, N_B molecules of gas B alone, and $(N_A + N_B)$ molecules of gases A and B together. Then, if the temperature T is the same in all three cases, eqn (2.3) tells us that

$$P_A V = N_A kT, \quad \text{(i)}$$
$$P_B V = N_B kT. \quad \text{(ii)}$$

Adding (i) and (ii), $(P_A + P_B)V = (N_A + N_B)kT$ (iii)

But eqn (2.3) also states that $P_{AB}V = (N_A + N_B)kT$ (iv)

Comparing (iii) and (iv) we see that $P_{AB} = P_A + P_B$, that is the pressure of a mixture of gases is the sum of the pressures due to each separately occupying the same volume. This is Dalton's law of partial pressures.

Kinetic theory, like all physical theories, is only a framework on which to hang experimental results. If the results don't fit, the theory must be modified. In fact most real gases obey eqn (2.4) quite well, though we have derived it for the very unreal case of an *ideal gas* composed of infinitely small, elastic, spherical molecules. More refined versions of kinetic theory take into account, for instance, the fact that real molecules have a finite size; thus the volume actually available to one of them is not V but a lesser amount, smaller than V by the total volume of the other molecules. Thus V is replaced by $(V - b)$ where b is this small correction. For further discussion of how one deals with these deviations from the ideal gas laws, see the book by Tabor (p. 219).

Another vital difference is that real molecules may possess internal structure. Only the noble gases are monatomic, with truly spherical molecules; diatomic molecules such as O_2 are dumbell-shaped, triatomic ones such as O_3 are triangular, and so on for more complex molecules. The atoms of such molecules can then move relative to the overall centre of mass, for instance diatomic molecules can vibrate back and forth along their axis, triatomic molecules can vibrate, and also rotate around all three axes, and so on. Such rotations and vibrations are a form of molecular kinetic energy distinct from the translational kinetic energy of the centre of mass. There is a remarkably simple relation between the complexity of a molecule and its (average) total kinetic energy, which we shall illustrate by discussing a special case. Consider a diatomic gas partially dissociated into its separate atoms (a situation which prevails to some extent in all gases). The atoms and molecules will be continually colliding with one another, and the effect of these collisions will be to equalize the velocity of the atoms and of the molecules. (This statement can be made plausible, without proving it. If one object overtakes and collides with another, then whatever their relative masses the faster will be slowed down and the slower speeded up. So after a period of time in which many such collisions occur, the atoms and molecules will eventually end up with the same average velocity, even though they might have had very different velocity distributions initially.) Since the average velocities are thus the same, the average energy of the molecules will be just twice that of the atoms; and since the latter is equal to $\frac{3}{2}kT$, the average energy of the diatomic molecules is $3kT$. Similarly we can show that an n-atomic molecule posseses $n \times \frac{3}{2}kT$ of kinetic energy. This result is a special case of the general principle of *equipartition*, which states that energy is shared equally, on average, amongst the various possible types of motion (sometimes called *degrees of freedom*) of an atom, molecule or similar microscopic object, each degree of freedom contributing $\frac{1}{2}kT$ to the total energy. Thus, a triatomic molecule can rotate about three independent axes, and the atoms can also vibrate to and fro along the three axes joining them, in addition to the whole molecule moving along three independent axes in space. The molecule thus has three rotational, three vibrational and three translational degrees of freedom, and so its energy is $9/2kT$.

Brownian motion

The principle of equipartition applies to any object in equilibrium* with its surroundings, no matter how large its mass. Thus one can calculate the average kinetic energy, and hence the mean velocity, of a macroscopic object in thermal equilibrium with its surroundings.

The Molecular Nature of Matter

Taking, for instance, a 40 g table-tennis ball in air at 300°K, we find on substituting in eqn (2.2) (we assume that the rotational and vibrational motion of the ball is negligible) that the mean velocity is about 5×10^{-10} m s^{-1} which is evidently virtually unobservable.

In 1827 Robert Brown, a Scottish botanist, first observed this thermal motion in pollen grains that he was studying in a water suspension under the microscope. Thinking that it might be evidence of some 'life force' he searched for it also in the water inclusions, containing mineral grains, found in certain ancient rocks. These mineral grains also exhibited the random jiggling motion, which Brown correctly attributed to the constant bombardment from all sides by the surrounding water molecules, and which is now called *Brownian motion*.

Example: A pollen grain 1 μm in diameter and of relative density 2 has a mass of about 10^{-15} kg. Hence

$$\overline{v^2} = \frac{3kT}{m} = \frac{3 \times (1 \cdot 38 \times 10^{-23}) \times 300}{10^{-15}} \simeq 12 \times 10^{-6} \text{ m}^2 \text{sec}^{-2}$$

and so $\bar{v} \simeq 3 \cdot 5 \times 10^{-3}$ m sec^{-1} $\simeq$ 3·5 mm sec^{-1}

In fact the grains do not appear to move as fast as this example would suggest because they are continually changing direction as other molecules collide with them. In a liquid, such as water, the rate of collision is typically 10^{12} per second. What we actually observe is the average motion as a result of all these collisions. It is therefore interesting to calculate the average distance from its starting point that such a particle will travel in a given time. This problem, commonly known for obvious reasons as the 'drunken walk problem', was solved by Einstein. We shall give a much simplified version of his solution. Then, having solved the problem for a macroscopic object, we shall apply the same sort of analysis to molecular motion (pp. 47–52). The reader may wish to skip this proof on first reading, and take the result (eqn 2.5) on faith.

What happens is that the particle takes a series of steps of varying length, each step making some indeterminate angle θ with the

---

* The concept of equilibrium is very important in physics. Roughly speaking a system is said to be in a *state of equilibrium* if no further changes occur when the system concerned is isolated from all other systems. Thus, the system consisting of a cup of hot coffee plus its cooler surroundings is not in equilibrium since the temperature of the coffee will decrease as time goes by. So a useful criterion for two objects' being in equilibrium with each other is that their temperatures be equal, and this is just another way of saying that their mean molecular kinetic energies are equal. This condition is known as *thermal equilibrium*.

X axis (see Fig. 2.2); θ is equally likely to have any value between 0 and 360°. Our first approximation is to assume that each particle travels the same distance λ between each collision, λ being known as the *mean (average) free path*. We also suppose that the particle is always travelling at a constant velocity equal to its average velocity $\bar{v}$. We will calculate the average squared distance, $\overline{x^2}$, travelled along the X axis in a time t seconds. Since the axes are arbitrary, $\overline{x^2} = \overline{y^2} = \overline{z^2}$ and so the total distance travelled, R, is given by Pythagoras' theorem;

$$\overline{R^2} = \overline{x^2} + \overline{y^2} + \overline{z^2} = 3\overline{x^2}, \text{ say.}$$

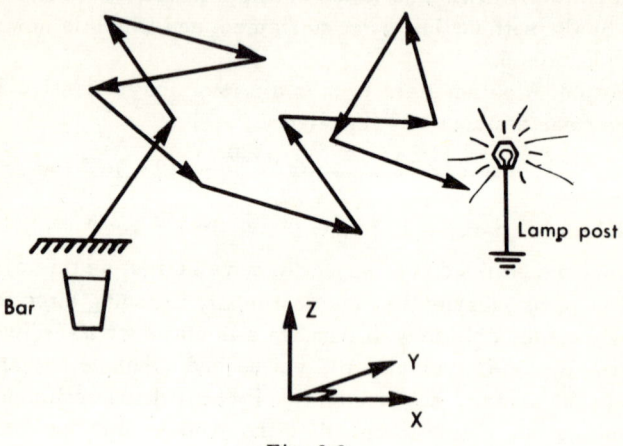

Fig. 2.2
The 'drunken walk' problem.

Now $\overline{x^2} = (\lambda \cos\theta_1 + \lambda \cos\theta_2 + \ldots)^2 = [\Sigma(\lambda \cos\theta_i)]^2$ where each angle θ is given a suffix i, and the sigma sign Σ tells us to sum the projected distances $\lambda \cos\theta_i$ over all the paths travelled in time t. If we carry out explicitly the process of squaring we find

$$\overline{x^2} = \Sigma(\lambda \cos\theta_i)^2 + 2\Sigma \lambda^2 (\cos\theta_i)(\cos\theta_j)$$

where the second term is analogous to the term $2ab$ in the expression $(a + b)^2 = a^2 + b^2 + 2ab$.

Now in one second, say, there will be about 10^{12} collisions so θ is going to have such an enormous number of different values that, on average, there will most certainly be as many in the range 270°–90° as there are in the range 90°–270°. These ranges correspond to $\cos\theta$ being respectively positive and negative,

The Molecular Nature of Matter

consequently $(\cos \theta_i)(\cos \theta_j)$ is equally as likely to be positive as negative, and so when we add up an enormous number of such terms we are going to get an answer which is effectively zero. That is, the second term summation can be neglected and

$$\overline{x^2} = \Sigma(\lambda \cos \theta_i)^2$$
$$= \lambda^2 \Sigma(\cos \theta_i)^2$$
$$= \lambda^2 \, \overline{\cos^2 \theta_i} \times \text{(total number of steps in } t \text{ seconds)}$$
$$= \lambda^2 \times \frac{1}{3} \times \frac{t\overline{v}}{\lambda}$$
$$= \frac{\lambda \overline{v} t}{3}$$

$\left(\text{since the average value of } \cos^2 \theta \text{ can easily be shown to equal } \frac{1}{3}\right).$

Hence finally
$$\overline{R^2} = 3\overline{x^2} = \lambda \overline{v} t \qquad (2.5)$$

The important feature of Brownian motion, shown by this equation, is that *the distance travelled is proportional to the square root of the time taken*. This property characterises the molecular motion known as *diffusion*, of which Brownian motion is a special case.

Example: Consider our 1 μm pollen grain, moving with an average velocity $\overline{v} \simeq 3 \cdot 5 \times 10^{-3}$ m sec^{-1}. Its mean free path $\lambda \simeq 10^{-9}$ m (roughly the intermolecular spacing in water). Hence in one second $\overline{R^2} = 10^{-9} \times (3 \cdot 5 \times 10^{-3}) \times 1 = 3 \cdot 5 \times 10^{-12}$, i.e. $\overline{R} \simeq 2 \times 10^{-6}$ m = 2 μm. Thus the grain will move, on average, a distance of the same order as its diameter, in one second. This is the sort of movement that Brown saw through his microscope.

The mean free path and the size of molecules

The mean free path λ is evidently a very useful parameter in kinetic theory. How is it to be measured? In the case of liquids, as in the example above, it is of the same order as the molecular size, but the molecules of a gas might be expected to travel much further than this before colliding, because of the much greater intermolecular spacing; we shall show that this is so by deriving a simple relationship between the size of molecules, their concentration and their mean free path.

Imagine that a single molecule is shot into a thin slab of gas containing n such molecules per unit volume. We suppose that each molecule is spherical with some radius r; obviously if the centres of two molecules approach each other within a distance $2r$, the molecules will collide. Thus the track of our molecule through the slab of gas is surrounded by a cylinder of radius $2r$, such that if the centre of any molecule lies within this cylinder there will be a collision. The cross-sectional area of this cylinder is $\pi(2r)^2 = 4\pi r^2 = 4\sigma$ where σ is the cross-section of a single molecule. The number of molecules that lie within the cylinder is simply n times the volume of the cylinder, that is $n \times 4\sigma \times d$ where d is the length of the cylinder, that is the slab thickness. If we now define the mean free path λ as the average distance between collisions, there will be one collision for a thickness, d, equal to λ, i.e.

$$1 = 4n\lambda\sigma \qquad (2.6)$$

(Note that, because of the rather simple-minded way in which we derived it the numerical factor 4 in eqn (2.6) is not quite accurate; however, the results obtained with its use are not very far from the experimental truth.)

Thus to determine λ we need to know σ, the molecular cross-section. We shall not discuss here the methods by which molecular sizes can be determined. They range from the conceptually simple (seeing how thinly one can beat a gold foil, or measuring the area to which an oil-drop on water can spread) to modern techniques of electron and X-ray scattering. Since molecules are really not hard spheres but rather fuzzy assemblages of electrically-charged particles, it is important to clarify what 'size' is measured by each method, and relate it to what is needed for a particular application. Since the kinetic theory is concerned mainly with elastic collisions between molecules, the size we want is the distance to which two molecules can approach one another without their internal structure being affected (e.g. without a chemical reaction occurring). This size is determined by the radius of the orbits in which the outermost electrons move (see Chapter 6) which is about 2·5 Ångström units ($2·5 \times 10^{-7}$ mm) for the H_2 molecule, and only a little larger for most simple molecules such as N_2, H_2O etc. On the other hand complex organic molecules such as proteins may have dimensions measured in hundreds of Ångström units. Now all gases at STP contain $6·02 \times 10^{23}$ molecules per 22·4 litres, that is n is about 3×10^{16} molecules mm^{-3}, and so for such a gas λ is given by

The Molecular Nature of Matter

$$\lambda \simeq \frac{1}{4\sigma n} \simeq \frac{1}{4\pi (2 \times 10^{-7})^2 \times 3 \times 10^{16}}$$

$$\simeq 7 \times 10^{-5} \text{ mm}$$

The average intermolecular spacing in such a gas is the reciprocal of the cube root of the concentration, that is about $0 \cdot 3 \times 10^{-5}$ mm. Clearly λ is very much larger than this, as stated above.

Diffusion

Diffusion is the process whereby molecules move from regions of high concentration to those of lower concentration. Like Brownian motion it is a random process; one cannot predict with certainty the path of a particular molecule, but the overall drift is exactly calculable. Observing that the rate of diffusion is greater the larger the difference in concentration involved, Fick in 1855 stated the fundamental law of diffusion, that the rate at which molecules cross unit area of a plane perpendicular to the direction of concentration difference (that is the *flux* or *current density*, J) is proportional to the rate of change of concentration with distance (that is, the concentration gradient). We can thus write

$$J = D \frac{\Delta n}{\Delta x} \tag{2.7}$$

where D is the constant of proportionality (the *diffusion constant*) and $\frac{\Delta n}{\Delta x}$ is the usual symbolic way of indicating a rate of change, the concentration n changing by a very small amount, Δn, in a small distance Δx. Now it happens that the diffusion constant can be calculated in terms of the mean free path λ and the mean molecular velocity $\bar{v}$. We shall show a simple way of doing this.

Imagine (Fig. 2.3) three parallel planes drawn in the gas, at right angles to the concentration gradient and at a distance apart equal to the mean free path, λ, so that each molecule on average travels from one plane to the next without undergoing a further collision. If we let the concentration at the central plane be n per unit volume, the concentrations at the left and right planes will be $\left(n + \lambda \frac{\Delta n}{\Delta x}\right)$ and $\left(n - \lambda \frac{\Delta n}{\Delta x}\right)$, respectively. Now consider how many molecules will cross unit area of the central plane per second, along the direction of the concentration gradient. Since the concentration

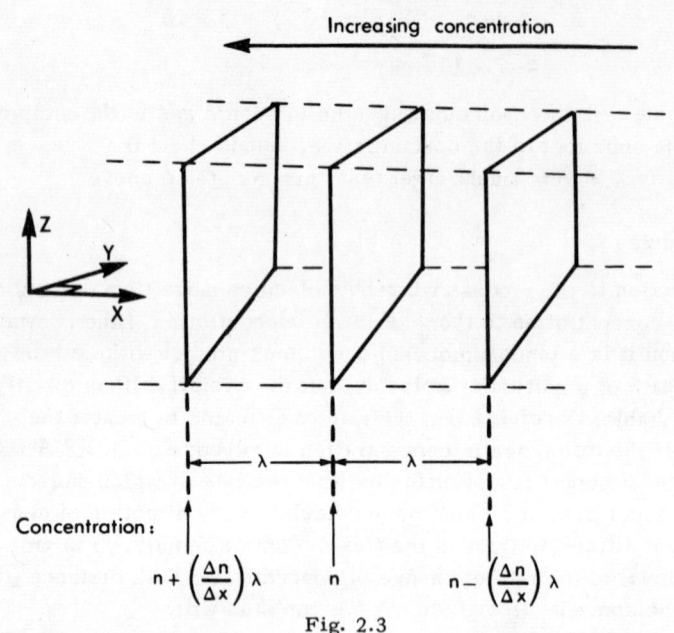

Fig. 2.3
Three parallel planes in a gas, perpendicular to the direction of increasing concentration.

is higher at the left than at the right, more molecules will pass from left to right than vice versa. Explicitly we assume that all molecules have the same velocity $\bar{v}$, and that one-sixth of them are travelling in each of the six independent directions $\pm X$, $\pm Y$, $\pm Z$. Then the number of molecules crossing unit area of the central plane per second, from left to right, is $\frac{1}{6}\left(n + \lambda \frac{\Delta n}{\Delta x}\right)\bar{v}$, and the number crossing from right to left is $\frac{1}{6}\left(n - \lambda \frac{\Delta n}{\Delta x}\right)\bar{v}$. The net flux is the difference of these two, that is

$$J = \frac{1}{6}\bar{v}\left(n + \lambda \frac{\Delta n}{\Delta x}\right) - \frac{1}{6}\bar{v}\left(n - \lambda \frac{\Delta n}{\Delta x}\right)$$

$$= \frac{1}{3}\lambda \bar{v} \frac{\Delta n}{\Delta x} \qquad (2.8)$$

By comparison with eqn (2.7) we see that the diffusion constant is given by

The Molecular Nature of Matter 49

$$D = \frac{\lambda \bar{v}}{3} \quad (2.9)$$

and so eqn (2.5) for Brownian motion can be written

$$\overline{R^2} = 3Dt \quad (2.10)$$

This emphasises the intimate relationship between Brownian motion and diffusion. The difference between the two concepts is this; whereas in the derivation of eqn (2.8) we assumed a constant concentration gradient along some particular direction (the X axis in our case) and calculated the rate of flow of molecules *in that direction*, eqns (2.5) or (2.10) tell us how far on average a particular molecule (or macroscopic object) will diffuse in a given time, *regardless of direction*, and independently of whether or not there is a concentration gradient in that direction. Let us illustrate the use of these equations by a specific example, the diffusion of gases, particularly oxygen and carbon dioxide, into and out of the tracheae of insects. The tracheae are the narrow tubes which lead from an insect's surface to its internal circulatory system, and which by diffusion carry oxygen inwards from the atmosphere and take the waste products of respiration outwards. Though, by mechanical movements such as wing beating, some insects can improve the efficiency of this process, the basic limitation on respiratory activity is set by the maximum possible diffusion rate, which is determined by the diffusion constant D and the dimensions of the tracheae. For oxygen at 20°C, eqn (2.2) tells us that $\bar{v} \simeq 4 \cdot 5 \times 10^5$ mm s^{-1}, and we have shown above that $\lambda \simeq 7 \times 10^{-5}$ mm. Let us consider a trachea of diameter 0·05 mm and length 2 mm, not untypical dimensions. We wish to calculate the flux J of oxygen molecules down the trachea, and the average time taken by these molecules on their journey. The relevant concentrations are the atmospheric concentration of oxygen externally and some fraction of this, typically one-half, internally. The concentration difference is thus about one-half atmospheric concentration, that is about

$\frac{1}{2} \times \frac{\frac{1}{5} \times 6 \times 10^{23} \text{ molecules}}{22 \cdot 4 \text{ litres}} \simeq 2 \cdot 5 \times 10^{15}$ molecules mm^{-3}. Substitution in eqn (2.8) yields

$$J = \frac{1}{3} \times (7 \times 10^{-5}) \times (4 \cdot 5 \times 10^5) \times \frac{2 \cdot 5 \times 10^{15}}{2}$$

$$\approx 10^{16} \text{ molecules mm}^{-2} \text{ sec}^{-1}$$

Since the trachea has a cross-sectional area of $\pi \left(\frac{0{\cdot}05}{2}\right)^2 \simeq 2 \times 10^{-3}$ mm², the total number of molecules respired per second, for each trachea, is $10^{16} \times (2 \times 10^{-3}) \simeq 2 \times 10^{13}$ molecules per second. The average molecular concentration during the journey down the trachea is three-quarters of atmospheric concentration, that is about 4×10^{15} molecules mm⁻³, so a volume flow of $\left(\frac{2 \times 10^{13}}{4 \times 10^{15}}\right) = 5 \times 10^{-3}$ cubic millimetres per second is occurring. Since the volume of the trachea is $2 \times (2 \times 10^{-3}) = 4 \times 10^{-3}$ mm³, each molecule respired has taken about $\frac{4 \times 10^{-3}}{5 \times 10^{-3}}$ seconds, that is about one second, to diffuse down the trachea.

This particular example has important consequences for insect life. Note that the length of the trachea occurs twice in the calculation, first in finding J, and second in determining the volume of the trachea. Thus the time taken is proportional to the square of the trachea's length. (Another way of seeing this is from eqn (2.10); although this applies to random diffusion, and the calculated time to diffuse a particular distance will be much less than we have just calculated, the fact that the time increases as the square of the distance involved is apparent.) The consequence for an insect with trachea ten times longer, say, is that oxygen will take about one hundred times longer to diffuse, and the resulting 'respiration time' of about one hundred seconds obviously puts such an insect at a severe disadvantage with respect to smaller, more rapidly-respiring ones! This is an important reason (there are others) why few insects have evolved to bodily dimensions very much greater than a few millimetres or so.

Two further general consequences follow from eqn (2.8). For a given concentration gradient the flux is proportional both to $\bar{v}$ and to λ. At constant temperature the product $(\tfrac{1}{2}m\bar{v^2})$ is constant, so $\bar{v}$ is inversely proportional to the square root of the molecular mass, m. That is, for two gases A and B with the same mean free path, the diffusion rates (fluxes) will be in the ratio

$$\frac{J_A}{J_B} = \frac{\bar{v}_A}{\bar{v}_B} = \sqrt{\frac{m_B}{m_A}}$$

This statement is sometimes known as Graham's law of diffusion. It implies that lighter gases diffuse faster; for instance hydrogen

The Molecular Nature of Matter

will diffuse at a rate $\sqrt{\frac{32}{2}}$ = 4 times faster than oxygen.

The second consequence of eqn (2.8) is that a reduction in the mean free path lowers the diffusion rate. This is not particularly relevant to gases, since we have already shown that at a particular concentration the mean free paths in all gases are very similar. However it becomes important when the kinetic theory is extended (as we have done implicitly already, in discussing Brownian motion) to include liquids as well as gases. A large number of chemical reactions in living organisms occur as a result of diffusion of reactants across cell-walls or through plasmas, and in these cases the mean free path may be much lower than that in gases, since the molecules are not only much more densely packed, but also generally larger than gas molecules.

Example: Compare the diffusion constant of oxygen in air with that of sucrose in aqueous solution. For oxygen, $D_{oxygen} \approx \frac{1}{3} \times (7 \times 10^{-5}) \times (4 \cdot 5 \times 10^{5}) = 10$ mm² sec⁻¹. For the sucrose solution, the mean molecular velocity $\bar{v}$ is less by the square root of the ratio of molecular masses, that is by $\sqrt{\frac{340}{32}} \approx 3$, and $\bar{\lambda}$ is less by about a factor of eight thousand (since σ is larger by a factor five and the concentration is roughly sixteen hundred times greater). Thus $D_{sucrose}$ is $\frac{D_{oxygen}}{3 \times 8000} \simeq 4 \times 10^{-4}$ mm² sec⁻¹. (The measured value is $5 \cdot 2 \times 10^{-4}$ mm² sec⁻¹.)

Fortunately for life, the distances involved in these processes are usually small enough that the reactions still occur very rapidly. For instance if sucrose molecules are required to diffuse across a cell wall 100 Å (= 10^{-5} mm) thick before a reaction can take place, the time taken can be estimated from eqn (2.10) as

$t \approx \frac{R^2}{3D} = \frac{10^{-10}}{3 \times (3 \times 10^{-4})}$ seconds, or about one-tenth of a microsecond.

When discussing diffusion through a thin membrane it is often convenient to introduce the idea of *permeability*. In a real-life situation where the diffusion equation, $J = D\frac{\Delta n}{\Delta x}$, is used, it is generally true that the most difficult quantity to measure is the distance Δx involved in the process. This is particularly so when the thickness of a dividing membrane is small and indeterminate or where diffusion occurs across some ill-defined boundary between two fluids. In an attempt to avoid this, the permeability P is defined as

$$P = \frac{D}{\Delta x}$$

and the diffusion equation becomes

$$J = P \, \Delta n \qquad (2.11)$$

The permeability of a process can be determined directly by measurements of molecular flux and concentration differences; thus P is more closely connected to experimental situations than is D, whose definition involves microscopic quantities such as λ and $\bar{v}$.

As an example, sucrose has a permeability with respect to an aqueous boundary of dimension, let us say, 10^{-2} mm (the interface involved must always be specified when stating permeabilites) of $\frac{D}{\Delta x} \approx \frac{5 \times 10^{-4}}{10^{-2}} = 5 \cdot 10^{-2}$ mm sec^{-1}. Thus permeability is expressed in the same units as velocity; it measures the rate of diffusional flow per unit concentration difference.

Biological tissues are characterised by extremely high permeabilities to certain ions and molecules, usually higher than can be attained with artificial interfaces. Also, substances of greatest biological interest are usually ionic in character, and for these the presence of voltage differences across a membrane leads to further complications. In general, these polar molecules have much lower permeabilities than do non-polar molecules such as sucrose. Taking these two factors together it is apparent that a theory to predict permeabilities will be difficult to construct, and in fact one always relies on experimentally measured values of P. Table 2.1 lists some of these values, for various molecular species passing through two typical* biological membranes.

Osmosis

One fact apparent from Table 2.1 is that biological membranes are extremely permeable to water (and alcohol), much more so than to other polar molecules. This property is probably due to the membran

---

* Note that some biological membranes are directional, and differentiate between cations and anions; this is a reflection of the electrical properties of the membrane and has important consequences for energy-exchange processes in cells.

The Molecular Nature of Matter

TABLE 2.1

Molecular species	Interface	Permeability (in mm sec^{-1})
K^+	Nerve axon of giant squid	4×10^{-6} (inwards and outwards)
Na^+		3×10^{-7} (inwards)
		4×10^{-5} (outwards)
Cl^-		3×10^{-7} (inwards and outwards)
Glucose	Wall of red blood cell	$< 10^{-9}$
Urea		6×10^{-6}
Water		3×10^{-3}

components being in a hydrated state, and is of immense significance. Such selectively permeable membranes are known as *semipermeable membranes* or SPMs, to indicate their marked preference for a diffusional transport of some molecular species rather than others. In particular, where water is the preferentially diffused molecule, we refer to the process as being one of *osmosis*. Osmosis is therefore *the diffusion of water from a place of high water concentration to one of lower water concentration, through an SPM*. There are many commonplace examples of osmosis in action. When dried fruits such as prunes or sultanas are soaked in water, the fruit skin acts as an SPM and the water diffuses from the outside (high water concentration) to the inside (low water concentration), swelling the fruit. The sultana, previously *flaccid*, becomes *turgid*. Wilting plants may be revived by watering; the root hairs have a semi-permeable surface layer which allows osmosis to proceed, and the normal water uptake to commence. Once in the plant cells the water stretches them, as the sultana is stretched, until they become rigid and the plant is erect.

In the latter example the process stops at some point, since the cell walls of living plants consist of strong cellulose which is able to resist any further stretching. However it is possible to burst prunes by leaving them to soak for too long in water. Diffusion will continue until either the concentrations of water inside and outside are equal, or until the pressure exerted by the membrane is sufficient to prevent more water entering. If the membrane is insufficiently strong it will break before this pressure is reached. Evidently therefore osmosis is associated with an *osmotic pressure*; we shall derive a theoretical expression for this and then see how it can be measured experimentally.

The technique we use is to draw an analogy between the behaviour of a solute in dilute solution and that of molecules in an ideal gas. The solute molecules, like the gas molecules, have a volume very small compared with the space they occupy, and collide with one another only occasionally. The only essential difference between the two situations is that the solute has a much smaller mean free path, because of the presence of the solvent molecules. However, the ideal gas law (eqn 2.4) is true whatever the value of the mean free path since it involves only the mean molecular velocities. Thus it seems reasonable that those laws that hold for an ideal gas should hold also for a solute in solution. In particular we shall adapt both the gas law and Dalton's law of partial pressures to derive the osmotic pressure of an aqueous solution.

Imagine a container divided into two parts by an impermeable membrane. Let one compartment be filled with water and the other be filled with an aqueous solution X. Suppose now that the membrane becomes quite permeable so that both water and X molecules can pass freely through it; as a result of diffusion it will be only a matter of time before the concentration of X has become the same in both compartments. If however the membrane were only semipermeable, passing water molecules but not X molecules, then the water would diffuse until its partial pressures in each compartment were equal; that is, the total pressure in the compartment containing solution would be greater than in that containing pure water by the partial pressure of the solute X. This partial pressure is the osmotic pressure (o.p.) of X, and it is usually denoted by the symbol π to distinguish it from the pressure of an ideal gas. Writing the gas law in terms of osmotic pressure, therefore, we have,

$$\pi V = nRT \qquad (2.12)$$

where n is the number of moles of solute in a volume V of solution.

Now we know that at standard temperature and pressure one mole of a gas occupies 22·4 litres. Thus one mole of solute in 22·4 litres of solution will exert an o.p. of one atmosphere, that is 760 mm of mercury or about 10^5 newtons per square metre. Remembering the definition of a *normal* solution as one containing one equivalent weight of solute per litre of solution, we see that the o.p. of a normal solution will be 22·4 atmospheres. This pressure is of course extremely large and accounts for the fact that osmosis is such an important process in biological systems, where the normalities involved are often considerable.

The Molecular Nature of Matter

There are two reasons why these simple arguments do not hold for real solutions, the o.p. of which is usually somewhat greater than that predicted by eqn (2.12). First, if dissociation of the solute occurs, each ion will act as a separate solute species, contributing its own o.p. to the solution; the total o.p. will be the sum of the partial o.p. of each ion. Thus, for an ionic solute such as NaCl, in which virtually complete dissociation occurs at all reasonably low concentrations, the o.p. will be twice that predicted by eqn (2.12), with intermediate values corresponding to varying degrees of dissociation. The second reason is this; eqn (2.12) applies to an 'ideal solution', that is one in which the solute molecules occupy a negligibly small volume and do not interact at all. Just as there are marked deviations from the gas law at sufficiently high pressures, so the o.p. of very concentrated solutions is somewhat higher than that predicted by eqn (2.12). For instance, the o.p. of a normal solution of sucrose is several atmospheres higher than 22·4 atmospheres.

For these reasons it is evident that measurements of o.p. made at non-zero concentration of solute will not provide a good check of eqn (2.12); to do this we must make measurements on an infinitely dilute solution, since only then can we be sure both that complete dissociation (of polar molecules) will have occurred and also that the solution will be 'ideal'. One way to do this in practice is to measure the o.p. of a solution at different concentrations, and then from a graph of $\left(\dfrac{\text{o.p.}}{\text{concentration}}\right)$ against concentration deduce the extrapolated value at zero concentration.

The o.p. of most substances can be measured fairly easily in the laboratory; the principle involved is in most cases that of measuring the external pressure that one has to apply to stop osmosis occurring. A simple method is to pour the solution into a tube or inverted thistle funnel, the lower end of which is closed by a flexible SPM such as the commonly-available films of cellulose acetate. The funnel is then immersed in water and the solution level rises as osmosis proceeds, stopping when the hydrostatic pressure exerted by the column of solution extending above the free water surface is equal to the o.p. of solution. Fig. 2.4 shows the results of some measurements made on sucrose solution using this method. The result of an extrapolation to zero concentration is also shown.

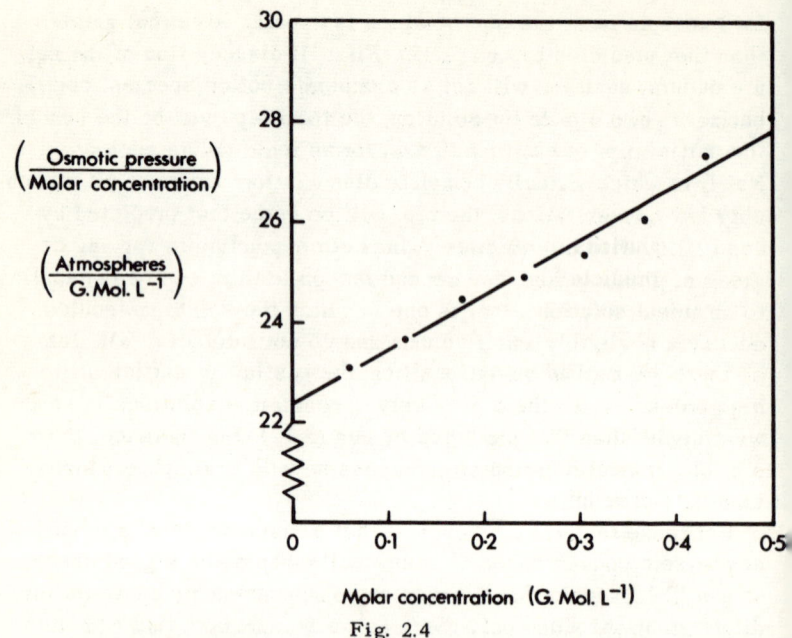

$\left(\dfrac{\text{Osmotic pressure}}{\text{Molar concentration}}\right)$

$\left(\dfrac{\text{Atmospheres}}{\text{G. Mol. L}^{-1}}\right)$

Molar concentration (G. Mol. L^{-1})

Fig. 2.4

Osmotic pressure per molar concentration of sucrose, showing the extrapolation to zero concentration.

Of course by this technique the o.p. measured is that of the final solution, after osmosis has stopped; if one requires to measure the o.p. of a given solution directly, one must exert an equal pressure on the other side of the SPM to stop osmosis occurring. This can be done fairly simply, for instance by using a mercury manometer on one side both to measure and to exert the pressure. (This is the principle of several forms of commercial osmometer.) If now the pressure applied is increased above that necessary to stop osmosis, water will be forced back through the SPM and the amount of water in solution will decrease at the same time as the amount of pure water on the other side of the SPM increases. This process, known as *reverse osmosis*, has become technologically interesting as a possible means of desalinating water, though so far problems such as the gradual deterioration of the available SPMs have prevented its application on a wide scale. (See the recent review articles referred to on p. 219.)

We end with two further examples of osmosis in action, considering first the case of red blood cells (erythrocytes) which float in

The Molecular Nature of Matter

blood plasma. Under normal healthy conditions the total concentration of ions (mainly Na^+ and Cl^-) inside the erythrocytes is about 300 milliequivalents per litre, compared with a concentration of similar ions of only 234 milliequivalents per litre in the surrounding plasma. These concentrations correspond to o.p. of 6·7 and 5·2 atmospheres, respectively. There is therefore a constant pressure difference across the erythrocyte wall of 1·5 atmospheres, quite enough to keep the erythrocyte fully distended. This water-balance is a delicate one; if the cell wall is ill-formed, or if various non-permeable species such as proteins occur in the wrong concentrations, the balance is upset and pathological damage may occur; one often attempts to treat such cases by means, such as the prescription of salt-free diets, which result in adjustments to the various ionic concentrations.

As a second example, osmosis is mainly responsible for water uptake by plants from the soil. It is not very easy to measure the solute concentrations in root-hairs of plants, relative to the soil, but an idea of the orders of magnitude involved can be got by considering the height to which water is forced in some plants. Trees can grow to heights of many metres, and water that reaches this height must be supported by a pressure of the order of an atmosphere, corresponding to a difference in solute concentration of about 50 milliequivalents per litre. (This is *not* the complete explanation of water take-up which is a complicated phenomenon involving a column of water in the plant stem being suspended under tension by transpiration pressure from the leaves; however it serves as an illustration of how important a process osmosis can be. In fact, in springtime the osmotic root-pressure is the dominant mechanism. See the bibliography for further reading on this topic.)

Sedimentation and the distribution of molecules in space

So far we have considered situations in which no forces act on molecules other than their mutual attraction or repulsion, and diffusion comes about as a consequence of the random nature of molecular collisions. Let us now turn to cases in which some force acts to urge all the molecules in a certain direction; we will ask what effect this force has on the diffusion process and on the distribution of molecules in space. We shall discuss in particular the force of gravity and the resultant effect of *sedimentation*, which, crudely speaking, may be defined as the piling up of molecules or larger objects at the bottom of a container. However, the argument used can be applied, with obvious modifications, to other forces such as electrostatic ones, to derive similar results. Though the arguments apply most readily to gases, the results can be extended as usual to include dilute solutions, or suspensions of solid particles in a liquid.

Consider therefore a large volume of gas (we may as well call it the 'atmosphere'), in which the force of gravity is pulling molecules down, whilst the randomizing effect of diffusion is tending to minimize the resultant increase in density at the bottom. We wish to find how rapidly the density falls off as we rise up in the atmosphere. To simplify the problem, in such a way that we can apply the gas laws directly, we shall assume that the temperature is everywhere the same in the gas, that is that the gas is *isothermal*;* such an idealized case is known as an *ideal atmosphere*. By the gas law (eqn 2.3) the pressure is given by

$$P = \frac{N}{V} kT$$

where there are N molecules in a volume V. If we denote the molecular concentration $\frac{N}{V}$, by n we have

$$P = nkT \tag{2.13}$$

Now on coming down in the atmosphere from height $(h + \Delta h)$ to height h the pressure increases from P to $(P + \Delta P)$, due to the weight of molecules in the intervening layer of gas. If we consider a vertical column of unit cross-sectional area this weight of molecules is the total number of molecules, $n \Delta h$, times the weight of each, mg.

Hence

$$\Delta P = -mgn \, \Delta h \tag{2.14}$$

the negative sign indicating that P decreases as h increases. But by letting both P and n in eqn (2.13) change by a small amount we see that

$$\Delta P = \Delta n \, kT \tag{2.15}$$

---

* The word 'isothermal' is also used to indicate *changes* in which the temperature stays constant, e.g. ice melts *isothermally* at 0°C.

The Molecular Nature of Matter

Equating the right-hand sides of eqns (2.14) and (2.15) yields

$$\frac{\Delta n}{\Delta h} = -\frac{mgn}{kT} \qquad (2.16)$$

By some mathematical manipulation it can be shown that eqn (2.16) is true if and only if the equation

$$n = n_0 \, e^{-\frac{mgh}{kT}} \qquad (2.17)$$

holds,* where n_0 is the concentration at height $h = 0$. Equations (2.16) and (2.17) are the fundamental equations of sedimentation.

The important point to note about eqn (2.17) is that the rate of fall-off of density depends on the molecular weight, being lower the smaller the value of m. So for example, one might expect the concentrations of oxygen ($m = 32$) and nitrogen ($m = 28$) in our atmosphere to diminish at different rates as the height increases. In fact our atmosphere is far from 'ideal' due to the great variation of temperature with height. This, combined with the stirring due to the winds, invalidates the conclusion completely for O_2 and N_2. However, hydrogen, which is present in microscopic yet detectable quantities at ground-level, has a sufficiently small mass for some residual effect to be noticeable; it is found that the molecular ratio of hydrogen to oxygen increases from two or three parts per million at ground level to several per cent at heights of 200 km or higher, and above about 1000 km hydrogen in fact predominates.

The term 'sedimentation' is commonly used to refer to the behaviour of solid particles suspended in a liquid. Equation (2.16) can be used to tell us whether or not particles in such a suspension will settle out, if it is left long enough to get into equilibrium. First we must decide what we mean by 'settle out'; our atmosphere decreases in density as we ascend but no one would say the molecules were 'settling out'. Since we wish to consider rigid, roughly spherical particles, let us say that they have settled out from the supernatant liquid if their concentration diminishes from a large to a small value in moving a distance equal to about one particle diameter, i.e. we require there to be a sharp boundary between particles and liquid. That is, we require $\frac{\Delta n}{n}$ to be $\geqslant 1$ (greater than, or of the order of, one) for $\Delta h \approx$ one diameter, d.

---

* This is one instance of the *exponential decrease* of a quantity, which derives from the nature of eqn (2.16). We shall meet further examples in Chapter 7.

Hence using eqn (2.16) the inequality

$$\frac{dmg}{kT} \geqslant 1$$

must be satisfied for settling-out to occur. The mass involved is of course the effective mass of the particle in the liquid, that is the mass of the particle minus the mass of liquid displaced.

Example: Consider red blood cells, suspended in the blood plasma. Will they settle out spontaneously? The diameter of red blood cells is about $6\,\mu$m, their thickness about $1\,\mu$m, and the difference between the density of the cells and of the plasma is about $0\cdot06$ gm cm^{-3}. Hence the effective mass of the cells is about 2×10^{-15} kg. Consequently, at 300°K,

$$\frac{dmg}{kT} \approx \frac{(6 \times 10^{-6})(2 \times 10^{-15}) \times 9\cdot81}{1\cdot38 \times 10^{-23} \times 300} \approx 30$$

and so this is obviously a case in which settling out will occur, if only after a considerable time.

It is often required to separate out particles which are so small, or whose density is so close to that of the liquid, that they will either not settle out naturally or will take an inconveniently long time to do so. In these cases we can increase the effective value of g by spinning the suspension in a *centrifuge*, which is simply a motor-driven turntable with arms to which cups can be attached. If the centrifuge spins with angular velocity ω radians per second, and the radius at which the particle is suspended is r, the 'centrifugal force' it experiences is (see Chapter 1) $mr\omega^2$. Comparing this with the gravitational force, mg, we see that the effective value of g is now $r\omega^2$. The ratio $\left(\dfrac{r\omega^2}{g}\right)$ is known as the relative centrifugal force, or RCF. Most simple student centrifuges can spin samples at a radius of about 10 cm, at 2000 revolutions per minute or so, giving an RCF of the order of several hundred. Specially constructed *ultracentrifuges* are used for research purposes and are capable of producing RCFs of hundreds of thousands.

We shall now extend this treatment of sedimentation under gravity to include all types of forces, and shall show how powerful are the results that can be obtained from this generalization. Equation (2.17) has the form

$$\frac{\text{Density at height } h}{\text{Density at height zero}} = e^{-\frac{(\text{potential energy at height } h)}{kT}}$$

The Molecular Nature of Matter

Although this equation was derived for the particular case of molecules moving under the influence of gravity in an ideal atmosphere, its general form is true for *all* forces, not only gravitational. That is, whenever the potential energy of atoms or molecules depends on their position, they will distribute themselves so that in the equilibrium state their densities are proportional to $e^{-\frac{(\text{potential energy})}{kT}}$. This is a very fundamental statement and has some very general and well-known consequences. For instance, at a particular temperature, T, the density will evidently be greatest when the potential energy is least. If therefore we plot a graph of potential energy against some coordinate representing the position, and find that the graph has a minimum for some value of the position coordinate, then the atoms or molecules will all tend to cluster around that particular position (just as, on the macroscopic scale, objects tend to roll downhill and concentrate in valleys). This situation is often true for real forces between atoms or molecules; at large distances apart they experience an attractive van der Waals' force, corresponding to negative potential energy, whereas at smaller distances the mutual repulsion of the electrically charged atoms increases rapidly, leading to a positive potential energy (see Fig. 2.5). Between the two extremes is a minimum of potential energy corresponding to the average distance apart of the atoms or molecules in a solid or liquid. If we know what this distance is, say by measuring the density of the material, then we can say something about the forces between atoms or molecules. Also, by studying how this equilibrium separation changes as the temperature is altered we can deduce the way in which the forces themselves, particularly the van der Waals' forces, vary as a function of temperature. Most common substances, for instance, expand with increasing temperature, and since the electrical forces do not vary very much with temperature we can say that the long-range attractive forces must be decreasing in strength (see Fig. 2.5).

Our next example is that of *evaporation*, and because of its biological importance we shall treat it in more detail. This is yet another case in which the laws of kinetic theory, derived for gases, are applied to molecules in liquids. As usual, our justification is not that we can deduce the laws rigorously from first principles (actually the study of liquids is one of the most difficult and least understood parts of physics) but that the results we obtain seem to tally very well with what is found experimentally to be the case.

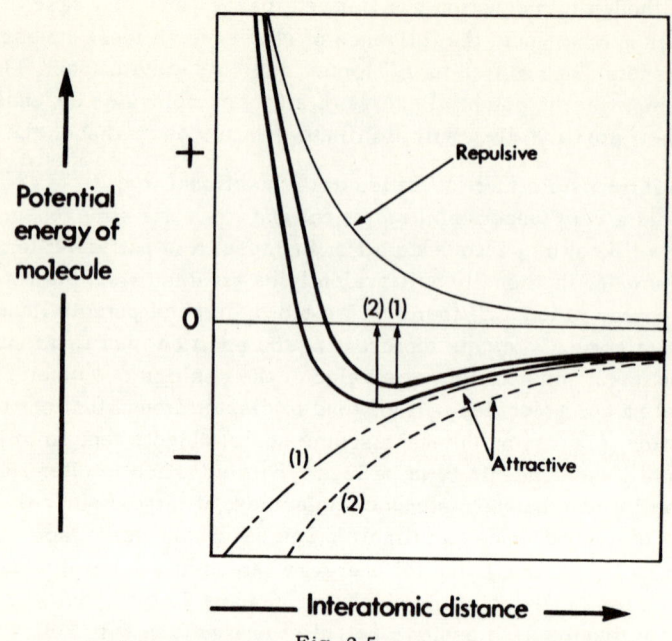

Fig. 2.5
The potential energy of a diatomic molecule as a function of interatomic spacing. We assume that there is a repulsive force (light line) between the atoms, and a variable attractive force (dashed lines). The heavy lines show the total resultant potential for two values, (1) and (2), of the attractive force. The corresponding equilibrium positions of the atoms occur at the minima of the total potential.

Consider then some liquid, containing N molecules per unit volume, with the surface in equilibrium with its vapour containing n molecules per unit volume. Let the potential energies of the molecules in the liquid phase and in the vapour phase be respectively E_l and E_v. We know E_v must be larger than E_l since energy has to be added to the liquid (by heating) to convert it to vapour. Knowing, therefore, that the densities of liquid and vapour are proportional to $e^{-\frac{(\text{potential energy})}{kT}}$ we can write the ratio of vapour to liquid density as

$$\frac{n}{N} = \frac{e^{-E_v/kT}}{e^{-E_l/kT}}$$

$$= e^{-(E_v - E_l)/kT}$$

$$= e^{-W/kT} \qquad (2.18)$$

The Molecular Nature of Matter

where $W = E_v - E_l$ and corresponds to the difference in potential energy of the molecules in the vapour and in the liquid. It is very instructive to work out a rough estimate for W, which we may call the *binding energy* of the molecules in the liquid relative to the vapour. In practice we convert liquid to vapour by boiling it, the latent heat being a measure of the energy needed to boil unit mass of a substance. Taking water as an example, its latent heat of vaporisation has been measured to be $2 \cdot 26 \times 10^6$ J kg^{-1}, and one kg of water contains $\left(\dfrac{6 \times 10^{23}}{0 \cdot 018}\right) = 3 \cdot 33 \times 10^{25}$ molecules. Hence one has to supply $\dfrac{2 \cdot 26 \times 10^6}{3 \cdot 33 \times 10^{25}} = 6 \cdot 8 \times 10^{-20}$ J completely to separate each molecule from its neighbours, and this is therefore a crude estimate of the binding energy W. Notice that the value of kT at room temperature, which we may take to be 300°K, is $1 \cdot 38 \times 10^{-23} \times 300 = 4 \cdot 14 \times 10^{-21}$ J, and so at this temperature the ratio $\dfrac{W}{kT}$ is $\dfrac{6 \cdot 8 \times 10^{-20}}{4 \cdot 14 \times 10^{-21}} = 16 \cdot 4$. So this simple example tells us that the molecular density of saturated water vapour at 300°K is only $e^{-16 \cdot 4}$ times that of the liquid. (The measured ratio of densities in fact corresponds to a value of about e^{-11}; this discrepancy arises because a lot of energy is also used to break up the ordered arrangement of molecules in the liquid, and to move them into random positions; we shall return to this energy of ordering in the next chapter, p. 80). A consequence of the result that W is very much greater than kT is that a small change in T has a very large effect on the vapour density relative to that of the liquid. For instance, let T change by 10%, say from 300°K to 330°K. Then in our example $\dfrac{W}{kT}$ will change from 16·4 to 18·0, corresponding to $\dfrac{n}{N}$ changing by a factor $e^{1 \cdot 6} \approx 5$. These values refer to equilibrium conditions, and to sustain them against vapour loss the liquid has to evaporate continuously. It should now be clear why a relatively small percentage increase in absolute temperature, say from arctic to hot desert conditions, can have a dramatic effect on the water-balance of living organisms. Note that we have assumed that the liquid and its vapour are in contact; the problem of water-loss has been overcome by many organisms, particularly plants, which interpose an impervious skin between their internal water-bearing organs and the atmosphere.

The Maxwell distribution

So far we have assumed that all molecules in a gas are moving with the same average velocity, $\bar{v}$. This simplification has enabled us to derive, correctly, a number of results that, had we adopted a more realistic view point, would have been difficult to prove rigorously. The fact is, however, that in an actual gas the molecules are moving with a whole spectrum of velocities, and there are certain branches of physics, for instance the study of the kinetics of chemical and biochemical reactions, where a knowledge of how the molecular velocities (and hence energies) are actually distributed is vitally important. The expression which gives the probability that a particular molecule will have a velocity between the values v and $v + \Delta v$ (or, equivalently, that gives the fraction of molecules with velocities between these limits) was derived by Maxwell in 1860 and the corresponding distribution of velocities is known as the Maxwell distribution. We shall not derive it rigorously but shall try to make it plausible.

Let us consider again the 'ideal atmosphere' of the last section. Molecules are moving in all directions, but to start with we consider only those moving up and down in the vertical direction, which we shall label the Z direction. From eqn (2.17) we know that the concentration of molecules at height h is

$$n_h = n_0 \, e^{-mgh/kT}$$

where n_0 is the concentration at ground level. We now ask, how many molecules which leave the ground and move vertically upwards will have sufficient energy to reach height h and contribute to n_h? The answer is that only those whose kinetic energy, $\frac{1}{2}mu^2$, is greater than the potential energy, mgh, can do so. If $\frac{1}{2}mu^2 = mgh$ the molecules will just arrive at height h, but with zero velocity. Consequently the total concentration at height h, n_h, is comprised of just those molecules which left ground level with vertical velocities greater than u.

$$n_h = n_0 \, (v_z > u)$$

where the term on the right means the number of molecules at height zero which have velocity greater than u in the Z direction. Substituting this into eqn (2.17) we have

$$n_0(v_z > u) = n_0 \, e^{-mgh/kT}$$
$$= n_0 \, e^{-\frac{1}{2}mu^2/kT}$$

This is not quite the result we want. It tells us the number of molecules with vertical velocities greater than a certain value, u,

The Molecular Nature of Matter

whereas we want the number in the range u to $u + \Delta u$. Fortunately it turns out that the algebraic expression for this is almost exactly the same, that is the probability of the vertical velocity's being in this range is

$$P(u) = C\, e^{-mu^2/2kT} \qquad (2.19)$$

where C is some constant of proportionality. We must choose C so that the probability is unity that the molecule has some velocity or other; C can then be shown to have the value $\sqrt{\dfrac{m}{2\pi kT}}$.

Equation (2.19) is the velocity distribution in one direction only; we really want the *speed* distribution, regardless of direction. Its derivation involves some difficult mathematics, and we just quote the result, which is

$$P(u) = 4\pi \left(\frac{m}{2\pi kT}\right)^{\frac{3}{2}} u^2\, e^{-mu^2/2kT} \qquad (2.20)$$

This is the *Maxwell distribution*. Apart from a trivial change in numerical constants, the important difference from eqn (2.19) is the introduction of a factor u^2, the effect of which is to make $P(u)$ go to zero as u gets very small. Consequently eqn (2.20) tells us that there is only an infinitesimal probability of finding a molecule at rest, with zero speed. This is reasonable, as although eqn (2.19) says that there is a finite chance of finding zero velocity *in a particular direction*, the chance of its being zero in all directions at once is obviously vanishingly small.

In Fig. 2.6 we sketch the distribution for two temperatures T_1 and T_2, showing how the probability of higher velocities increases with T. We also show the *most probable velocity*, equal to $\sqrt{\dfrac{2kT}{m}}$, for each curve.*

The results of this section and the last may be summarized by stating that the probability of a molecule having a particular velocity is given by $e^{-\frac{\text{kinetic energy}}{kT}}$, whilst the *density* (*concentration*) distribution is given by $e^{-\frac{\text{potential energy}}{kT}}$. We shall now apply these results to a simple chemical reaction, $A + B \rightleftharpoons AB$, to determine the factors which influence its rate and equilibrium position.

---

* Note that the most probable velocity is equal neither to the average velocity, $\bar{v}$, nor to the square root of the average velocity squared, $\sqrt{\bar{v^2}}$ (see discussion following eqn 2.1). However, for most practical purposes all these measures of velocity may be considered equivalent.

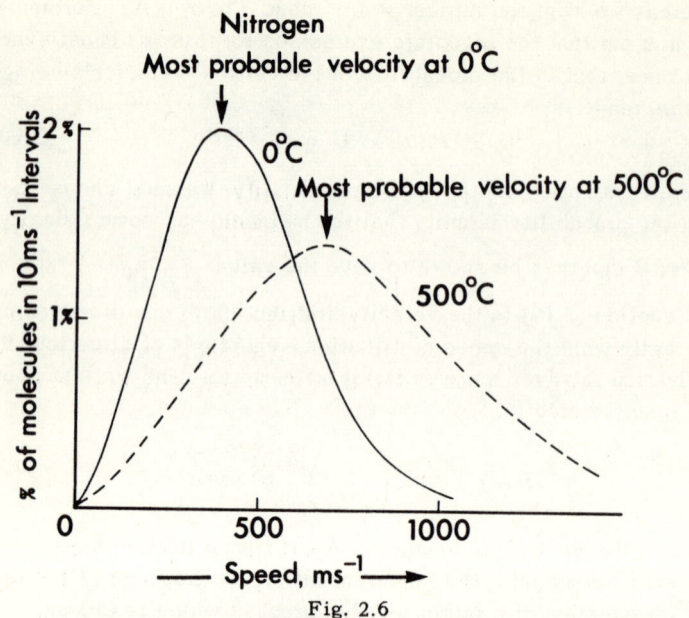

Fig. 2.6
The Maxwell distribution plotted for two temperatures T_1 and T_2. The most probable velocity, $\sqrt{2kT/m}$, occurs at the maximum of each curve.

Considering the latter first, our study of evaporation led to the conclusion of eqn (2.18), that the equilibrium concentrations of 'free' and 'bound' molecules were in the ratio $e^{-W/kT}$, where W is the binding energy. Similar arguments can be used for chemical reactions, and we deduce that the concentration of 'free' reactant A is $e^{-W/kT}$ times that of the product AB, that is

$$[A] = \frac{[AB]}{[B] \times V} e^{-W/kT}$$

(In accordance with chemical usage we now denote concentrations by square brackets.) Notice that the volume available for A atoms to be bound in the product AB is not unity, as for free A atoms, but is the number of free B atoms times some characteristic interaction volume, V. This accounts for the denominator $[B] \times V$ on the right-hand side. Hence the equilibrium concentrations are given by

$$\frac{[A][B]}{[AB]} = \frac{1}{V} \times e^{-W/kT} \qquad (2.21)$$

The Molecular Nature of Matter

and since V is to a good approximation independent of temperature it is clear that the *equilibrium* position of the reaction is determined by the temperature and the binding energy W of the product. The right-hand side of eqn (2.21) is often called the equilibrium constant. This however tells us nothing about the rate at which the reaction will proceed, starting from a mixture of A and B with no product present. What physically happens is that A and B atoms are constantly colliding, and it is true of most reactions that there is a large probability that they will not interact but will bounce off each other. Only if they possess sufficient energy to overcome their mutual repulsion at short distances (see Fig. 2.5) will an interaction occur. This energy, the *activation energy* E^*, determines the rate at which $A + B \rightarrow AB$ occurs, according to the expression

$$\text{Rate}_{A + B \rightarrow AB} = C_1 [A][B] e^{-E^*/kT} \qquad (2.22)$$

Similarly, the rate for the reverse reaction is given by

$$\text{Rate}_{AB \rightarrow A + B} = C_2 [AB] e^{-(W+E^*)/kT} \qquad (2.23)$$

(Note that at equilibrium these two rates must be equal, and equating the right-hand sides of eqns (2.22) and (2.23) we arrive back at eqn (2.21), as we must.) Therefore the *rate* at which the reactions occur depends not only on T and W, but also on E^*. By inspecting the Maxwell distribution one can determine what fraction of atoms will possess a velocity high enough for their energies to be above E^*.

Table 2.2 lists activation energies for some chemical reactions of biological importance. Note two points in particular about this table. First, the value of E^* for many reactions is about 5×10^4 J mol^{-1}, and comparing this with the value of RT (that is, kT evaluated for 1 mol of substance), which is $2 \cdot 47 \times 10^3$ J mol^{-1} at 300°K, we see that $\dfrac{E^*}{kT}$ has a value near 20 for many reactions. Consequently, as in our discussion of evaporation, it is clear that a small change in temperature leads to a large change in reaction rates. For the example just considered, a rise from 27°C to 37°C will increase the rate by the factor $e^{20 \times \frac{10}{300}}$, that is, by about a factor of two. The effect is even more pronounced for the last entry in the table, protein denaturation, the rate for which is diminished by twenty times or more when the temperature falls from 100°C to

TABLE 2.2

Process	Catalyst	Activation energy, E^* (J mol^{-1})
Free radical recombination	–	0
Inversion of sucrose	Acid solution	$8·6 \times 10^4$
	Malt invertase	$5·4 \times 10^4$
	Yeast invertase	$4·8 \times 10^4$
Decomposition of H_2O_2	–	$\sim 7·3 \times 10^4$
	Finely divided platinum	$4·8 \times 10^4$
	Liver catalase	$2·3 \times 10^4$
Hydrolysis of urea	Acid solution	$10·2 \times 10^4$
	Urease	$(2·7 - 5·2) \times 10^4$
Denaturation of protein		$\sim (10–60) \times 10^4$

90°C. This clearly explains the difficulty in boiling eggs (denaturation of egg albumen) at high altitudes where water boils below 100°C.

Second, note the column headed 'catalyst'. This shows that the activation energy for a reaction is not a constant, like the binding energy, but depends on the conditions under which the reaction occurs, and on the presence of other substances which probably act as intermediaries and may help to reduce E^*. Besides surface-acting catalysts, such as finely-divided platinum, and the catalysing role of acidic environments, we also stress the importance of biochemical catalysts, or *enzymes* (names ending in -ase). These are generally much more efficient at reducing E^* than are man-made substances, but their precise mode of action is often obscure. See the bibliography for an up-to-date reference to this topic.

More will be said about chemical reactions in the next chapter.

Viscosity

We end this chapter with a short discussion of viscosity and its application to the flow of blood. Although it would be possible to develop the subject by the methods we used earlier, that is, explaining large-scale effects in terms of the motions of molecules, it would be difficult and inappropriate to apply this technique to viscosity; difficult, because the physics involved is complex, and inappropriate because the results of greatest practical interest hold irrespective of the molecular nature of the fluid involved. We start, then, with the observation that the flow of fluids (liquids, gases)

The Molecular Nature of Matter

is never perfect, but is always retarded by internal frictional forces which tend to dissipate the kinetic energy of flow as heat. These forces are termed *viscous* and are due to the fact that, when layers of fluid are slipping past one another at different velocities, there is a continued breaking and re-making of 'links' between molecules in the various layers. Although the intermolecular forces in liquids and gases are not strong (at least, compared with those in solids) they are quite strong enough for viscous forces to dominate the flow of most fluids and to be the main problem met with in the design of aeroplanes, ships, pipe-lines and similar structures.

To discuss viscosity quantitatively we make the observation that, at the boundary between a moving fluid and a solid surface at rest, there is an (extremely thin) layer of fluid which is also at rest. For example, the blades of fans which have been running for some time become covered with a layer of very fine dust which is not blown off in spite of the relative motion of the blades through the air. This *boundary layer* exerts a viscous drag on the next fluid layer, and so on into the fluid, resulting in the velocity of the fluid flow increasing from zero at the surface to some value, let us say v, at a distance d away. We speak of a *velocity gradient*, $\frac{v}{d}$, existing in the fluid. Now imagine that the solid surface is a flat plate of area A; it is found experimentally that the force F required to keep the fluid flowing at constant velocity is proportional both to A and to $\frac{v}{d}$, that is

$$F = \eta \times A \times \frac{v}{d} \qquad (2.24)$$

where the constant of proportionality, η is measured in units of $\frac{\text{newton metres}}{(\text{metre}^2) \times (\text{metre sec}^{-1})}$ that is in $N\,s\,m^{-2}$ or $kg\,s^{-1}\,m^{-1}$. (In many places one will still come across an older unit, the *poise*, equal to one-tenth of the present unit.) Like other molecular properties viscosity is strongly temperature-dependent, though its behaviour is different for liquids and gases, the viscosity of liquids decreasing with temperature and that of gases increasing. (This is the first time we have found a pronounced difference in the kinetic behaviour of liquids and gases, and it reinforces the earlier statement that a rigorous treatment of viscosity is quite difficult.) Table 2.3 lists the viscosities of some fluids at various temperatures. We also list another quantity, the *specific viscosity*, defined as η

divided by the density, ρ, of the fluid. Its relevance is twofold; first, it is found empirically that although fluids differ greatly in their actual viscosity coefficients, their specific viscosities are often rather similar, as can be seen from the table. Secondly, it is found that the measured viscosity of a solute or suspension varies with solute (or suspension) concentration in a manner that differs from one solvent to the next. However, if instead the specific viscosity is measured and plotted against density of solution, a simple straight-line graph is obtained in nearly all cases. If this graph is extrapolated backwards to the point at which the density is equal to that of the pure solvent, a value for the specific viscosity is obtained which is independent of concentration and is determined only by the nature of the solute and solvent. (This is similar to the technique for finding osmotic pressures of infinitely dilute solutions, and is used in practise for molecular weight determinations, etc.; see Fig. 2.4.) It should be noted that solutions or suspensions usually have higher viscosities than the pure fluid — see the results for whole blood.

TABLE 2.3

Substance	Temperature	η (kg s^{-1} m^{-1})	$\frac{\eta}{\rho}$ (m^2 s^{-1})
Water	0°C	1·79 × 10^{-3}	1·79 × 10^{-6}
	37°C	0·69 × 10^{-3}	0·70 × 10^{-6}
	100°C	0·28 × 10^{-3}	0·29 × 10^{-6}
Mercury	0°C	1·685 × 10^{-3}	0·12 × 10^{-6}
	100°C	1·240 × 10^{-3}	0·09 × 10^{-6}
Air	0°C	1·71 × 10^{-5}	13·2 × 10^{-6}
	100°C	2·20 × 10^{-5}	23·1 × 10^{-6}
Blood plasma	37°C	1·18 × 10^{-3}	1·14 × 10^{-6}
Whole blood (20% cells, v/v)	37°C	∼1·7 × 10^{-3}	∼1·6 × 10^{-6}
Whole blood (60% cells, v/v)	37°C	∼5 × 10^{-3}	∼4·5 × 10^{-6}

Viscosity can be measured by many techniques, two of the more common being the falling-body method and the rate of flow method (the most used in practice). In the former a small sphere is dropped through the fluid, and its rate of fall measured. A simple equation (Stoke's law) gives the viscosity in terms of the sphere's radius, r, and its apparent density, ρ, as

$$\eta = \frac{2r^2}{9v} \rho g$$

The Molecular Nature of Matter

where v is the steady velocity at which the ball falls. This method is suitable only for transparent, viscous liquids available in large quantities. The viscosity of opaque liquids, gases, or fluids available only in small amounts is best found by measuring the rate of flow (volume per second) of fluid through a tube of radius r and length l when a pressure difference ΔP is applied across the ends. If V is the observed rate of flow, it can be shown that

$$\eta = \frac{\pi r^4 \Delta P}{8lV} \qquad (2.25)$$

(For suspensions, such as blood, the values found this way tend to be unreliable if the tube radius is comparable with the dimensions of the suspended particles.) Equation (2.25) is known as *Poiseuille's law*. The bulk flow of fluid for which it holds is often called *capillary flow*.

So far we have assumed that the flow is steady and smooth not turbulent. Steady flow is also called *streamline* flow since the paths of individual particles in the fluid are smooth curves, or streamlines. In *turbulent* flow, on the other hand, the streamlines break up into highly complex eddies and vortices, and since intermolecular bonds are being broken and remade much more often, the force required to drive a given quantity of fluid in turbulent flow is very much greater than in streamline flow. The transition is clearly seen in smoke rising from a cigarette, the initial smooth streamline flow soon breaking up into eddying turbulence. Turbulent flow is to be avoided in general, since so much energy is wasted in internal friction. How, then, can one tell when flow will become turbulent? The onset of turbulence in pipes is clearly indicated by injecting small quantities of dye into a fluid to mark the streamlines, and many experiments have shown that turbulence will not set in if the fluid velocity v is kept below a certain value which depends on the pipe diameter d and the fluid's density ρ and velocity η. It is possible to define a quantity, usually called Reynolds' number, $\mathcal{R}$, in the following way:

$$\mathcal{R} = \frac{\rho v d}{\eta}.$$

Putting in the relevant unit for each quantity in the equation, we see that $\mathcal{R}$ is a pure number: i.e.

$$\frac{(\text{kg m}^{-3})(\text{ms}^{-1})(\text{m})}{(\text{kg s}^{-1} \text{m}^{-1})}$$

is independent of the system of units used. The empirical fact is that turbulence is observed to set in for values of $\mathcal{R}$ exceeding approximately 2000, for *all* fluids. There is as yet no satisfactory explanation of this, but its importance is obvious.

Example: Crude oil of density 900 kg m^{-3} and viscosity 0·1 N s m^{-2} flows in a pipe of diameter 1 m. If turbulence is to be avoided, the oil velocity must be less than about $\dfrac{2000 \times 0·1}{900 \times 1} = \dfrac{2}{9}$ m s^{-1}, corresponding to a flow rate of $\dfrac{\pi d^2}{4} \times v = \dfrac{\pi}{4}(1)^2 \times \dfrac{2}{9} = 0·175$ m^3 s^{-1} = 157 kg s^{-1} = 0·157 tonnes per second.

The velocity of blood-flow can be measured in blood vessels of varying sizes, and it is a remarkable fact that, when the values obtained are substituted in the expression for Reynolds' number, a value very close to 1000 is always found. Moreover, laboratory measurements have shown that the viscosity of blood increases rapidly at velocities lower than this, corresponding to lower values of $\mathcal{R}$. (This behavior is characteristic only of suspensions, blood plasma itself behaving as a normal, so-called 'Newtonian' fluid whose viscosity is independent of flow-rate.) It appears therefore that our circulatory system is such that the heart does the minimum amount of work necessary to pump the blood at a given rate; attempts at higher or lower blood velocities would *increase* the work done by the heart, in one case through turbulence setting in and in the other through the rapidly increasing viscosity.

3

THERMODYNAMICS

This chapter is about energy. Specifically, thermodynamics is a topic concerned with energy transformations in systems containing a large number of particles. The results we shall obtain do not depend upon the particular details of energy transformation at the molecular or atomic level; a study of such details in biological systems is known as *bioenergetics*, and a textbook on this is cited in the bibliography. Rather, the results stem from an application of general laws of physics to systems containing many particles. Consequently thermodynamic laws have a wide range of applicability in physics, chemistry, biology, engineering — in fact in any discipline where energy transformations in large systems are relevant. If thermodynamic results depended upon particular molecular mechanisms this would clearly limit their area of usefulness. The results that we will discuss are based upon, and elaborate somewhat, the laws of mechanics and molecular motion set out in Chapters 1 and 2. The first result in fact is nothing more nor less than a restatement of the law of conservation of energy.

The first law of thermodynamics

The first law of thermodynamics states that the total energy of a system and its surroundings is a constant. The term *system* refers to a specific quantity of matter. It might, for example, refer to a mole of gas which was the subject of an experiment, or it could equally well refer to a living organism. The term *surroundings* refers to the rest of the universe, apart from the system. (Of course this is the most general possible definition of 'surroundings'; in actual practice it is rather unusual for a laboratory experiment to have much influence on the universe outside the laboratory, or for a frog in Europe to be much concerned with, or affected by, the weather in Australia. It is wise, however, to remember that in general the word 'surroundings' does not have such a restricted meaning.) A more common formulation of the first law is that the increase in internal energy of a system is equal to the heat supplied to the system plus the work done on the system by the surrounding. By *internal energy* of the system is meant the sum for all the molecules of the kinetic energy corresponding to their various degrees of freedom. We recall from Chapter 2 that an n-atomic molecule has $3n$ degrees of freedom, each with an average kinetic energy of $\frac{1}{2}kT$, where k is Boltzmann's constant and T the temperature in degrees kelvin. The first law can thus be written algebraically as

$$\Delta E = \Delta q + \Delta W \tag{3.1}$$

where ΔE, Δq and ΔW are the changes in internal energy, heat and work respectively. Naturally the same units must be used for each of the terms in this equation. For instance each could be expressed in joules. ΔW is taken to be positive if the surroundings work *on* the system and Δq is taken to be positive if heat is supplied *to* the system by the surroundings.

In order to illustrate how the first law is used we will consider the expansion of an ideal gas at constant temperature. An ideal gas, as discussed in Chapter 2, is composed of indefinitely small, elastic, spherical molecules. We will suppose that one mole of such a gas is enclosed in a cylinder, as in Fig. 2.1. The internal energy of the gas is simply the sum of the kinetic energies of all the molecules. The internal energy, since it is found by adding together terms of the form $\frac{1}{2}kT$, therefore depends only on the temperature and not, it must be stressed, on the pressure or volume, say, of the gas. Thus during such an expansion at constant temperature the change in internal energy, ΔE, is zero and hence

$$\Delta q = -\Delta W \tag{3.2}$$

Now ΔW will be negative because the system (the mole of ideal gas) has to do work on the surroundings as it pushes back the piston. Thus Δq is positive which means that heat must be supplied to the gas by its surroundings to maintain the constant temperature.

Later in this chapter it will be useful to have an expression for ΔW in terms of the initial and final volumes V_i and V_f and the constant temperature T. So although it is not strictly relevant to the

Thermodynamics

first law of thermodynamics we will derive the expression here. As we saw in Chapter 1 the work done by a gas in expanding its volume by a very small amount ΔV at pressure P is $P\Delta V$ (eqn 1.28). The equation governing the relation between pressure, volume and temperature (eqn 2.4) is

$$PV = RT \qquad (3.3)$$

This relation between P and V is represented by the smooth curve in Fig. 3.1. The work done in expanding the gas by a small amount ΔV at pressure P, $P\Delta V$, is the area of a strip AB below the curve. Thus the total work done is the sum of the areas of all the small strips like AB which lie under the curve between the initial and final volumes. Hence the total work done equals the area $CDEF$ shown in Fig. 3.1. The best way to calculate this area would be to use calculus,* and if this is done then it is found that

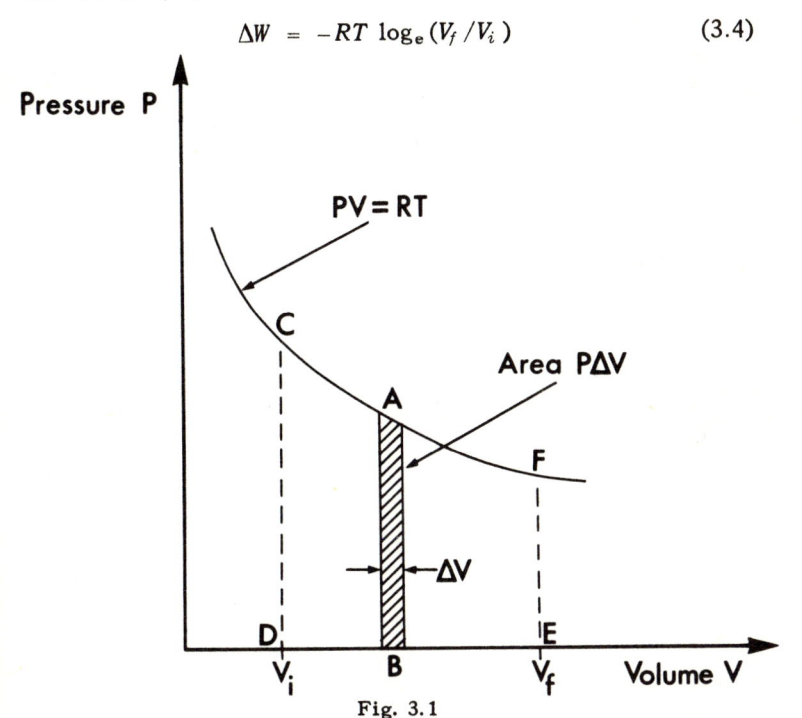

$$\Delta W = -RT \log_e (V_f / V_i) \qquad (3.4)$$

Fig. 3.1

* Another way in which this area can be found is by dividing it into a large number of strips like AB. Each strip is approximately rectangular. Assuming that they are precisely rectangular allows one to calculate the total area. The approximation gets better as the number of strips increases, and this is how such calculations are performed by computers.

Example: How much heat must be supplied to a mole of an ideal gas as it expands from an initial volume of 18 litres to a final volume of 20 litres, in order that its temperature may be maintained constant at 27°C? The heat supplied to the system is given by eqns (3.2) and (3.4)

$$\Delta q = -\Delta W$$
$$= RT \log_e (V_f / V_i)$$
$$= 8\cdot 31 \times 300 \times \log_e (20/18)$$
$$= 262 \text{ J}$$

The temperature T is in degrees kelvin and the numerical value of the universal gas constant R of $8\cdot 31$ J °K^{-1} has been used. Since there are 4·186 joules per calorie Δq is 62·6 calories.

To summarise this section we repeat that the first law of thermodynamics is a statement of the conservation of energy. Energy can take many different forms — electrical, chemical, kinetic, etc. — and during conversion from one form to another it is not possible to have a nett gain or loss of energy, when everything relevant to the system and its surroundings is taken into account. It is this law which shows that it is impossible to construct a machine that would constantly produce more energy than was put into it, i.e. a perpetual motion machine.

The first law, then, is a very general statement. It is so wide ranging, in fact, that one is often uncertain how to apply it in particular cases. It is common practice to introduce a subsidiary quantity called *enthalpy* which is defined in such a way that the first law can be applied directly to specific physical, chemical and biochemical systems.

Enthalpy

The enthalpy H of a system is defined as

$$H = E + PV \tag{3.5}$$

where E is the internal energy. We shall show soon that when enthalpy is defined in this way the heat released in a chemical or physical reaction taking place *at constant pressure* is equal to the change in enthalpy. The reason why enthalpy rather than internal energy is the relevant quantity is that chemical reactions, particularly in the gaseous phase, are in general accompanied by a volume change when they take place at a constant pressure. Work is done during this volume change and hence the heat released is *not*

Thermodynamics

exactly equal to the change in internal energy but to the change in enthalpy. (There is, however, a simplification which arises if the biochemical reactions of living organisms are considered. These reactions all take place in solution, gases, for example, being dissolved before they are used, and so any volume changes that do occur are negligible. Thus the reactions can be considered to take place at a constant volume, and the distinction between internal energy and enthalpy can be approximately ignored.)

We will now show that the change in enthalpy during a reaction is equal to the heat absorbed or released. Suppose that a small change in the variables E, P and V occurs. Call the amount by which each of the variables alters ΔE, ΔP, and ΔV respectively. The symbol Δ is used, as before, to indicate that the changes are small ones. The enthalpy of the system after this small change is given by

$$H + \Delta H = (E + \Delta E) + (P + \Delta P)(V + \Delta V) \quad (3.6)$$

The change in enthalpy is obtained by subtracting eqn (3.5) from eqn (3.6). This gives

$$\Delta H = \Delta E + P\Delta V + V\Delta P + \Delta P \Delta V \quad (3.7)$$

The term $\Delta P \Delta V$ can be dropped on the grounds that since it is the product of two small quantities it must itself be very small indeed and can be ignored in comparison with the other terms. The term $V\Delta P$ can also be dropped since we are considering a reaction occuring at a constant pressure and so $\Delta P = 0$. Finally $-P\Delta V$ is the work ΔW done by the system on the surroundings. Thus eqn (3.7) becomes

$$\Delta H = \Delta E - \Delta W \quad (3.8)$$

We can rewrite the right hand side of this equation using the first law of thermodynamics, eqn (3.1) This gives

$$\Delta H = \Delta q \quad (3.9)$$

where Δq is the heat evolved or absorbed. This proves the equivalence between change in enthalpy and amount of heat released.

Example: Consider the total combustion of stearic acid

$$C_{18}H_{36}O_{2(SOLID)} + 26\ O_{2(GAS)} \rightarrow 18\ CO_{2(GAS)} + 18\ H_2O_{(LIQUID)}$$

for which the change in internal energy per mole at 20°C is 2711·8 kilocalories. Calculate the heat evolved if the reaction is performed at constant pressure. The heat evolved Δq is equal to the change

in enthalpy ΔH, where
$$\Delta H = \Delta E + P\Delta V$$
$$= 2711 \cdot 8 + P\Delta V$$

Now the volume of the system is reduced during the reaction. Initially there are 26 moles of oxygen plus the volume of the solid stearic acid. At the end there are 18 moles of carbon dioxide plus 18 moles of liquid water. The volumes of the solid and liquid are both negligible compared with the gas volumes involved and so they will be ignored. Since the volume of the system is reduced the surroundings work on the system and the numerical value of $P\Delta V$, as we shall see, is positive. The gas law for n moles of gas is

$$PV = nRT \tag{3.10}$$

If in the course of the reaction V changes to $V + \Delta V$ and n changes to $n + \Delta n$ then

$$P(V + \Delta V) = (n + \Delta n)RT \tag{3.11}$$

Subtraction of eqn (3.10) from eqn (3.11) gives

$$P\Delta V = \Delta n\, RT \tag{3.12}$$

which is a positive quantity, as we said. Now initially there were 26 moles of gas, reducing to 18 moles in the final state. Thus $\Delta n = 8$ and so $P\Delta V = 8 \times 1 \cdot 987 \times 293$ calories $= 4656$ calories where a value of $1 \cdot 987$ calories per mole per degree kelvin has been used for the universal gas constant. Thus, finally

$$\Delta q = \Delta H = 2711 \cdot 8 + 4 \cdot 66 \text{ kilocalories}$$
$$= 2716 \cdot 5 \text{ kilocalories}$$

This was a rather straightforward application of the ideas of enthalpy. However it is difficult to determine the heat evolved in certain chemical reactions because they do not normally take place in an isolated system. An example is the difficulty encountered in measuring the heat of reaction for

$$C_6H_{12}O_6 \rightarrow 2\, C_3H_6O_3 \tag{3.13}$$

that is, glucose transforming to lactic acid, by direct calorimetry. The reason for the difficulty is that the reaction proceeds only in the presence of a complicated mixture of enzymes, coenzymes and salts and is accompanied by changes in pH and by phosphate transfer. We can resolve the difficulty if we know the energetics of the

Thermodynamics

breakdown (or the synthesis) of the glucose and the lactic acid separately. Now it is known that the heat of combustion of glucose is $\Delta H = -673$ kilocalories per mole and that of lactic acid is $\Delta H = -327$ kilocalories per mole. The end products of the total combustion of one mole of glucose and of two moles of lactic acid are identical (cf. eqn 3.13). Thus the enthalpy difference between one mole of glucose and two moles of lactic acid is equal to the difference between the enthalpy change due to total combustion of one mole of glucose and that due to the total combustion of two moles of lactic acid. Thus the heat of reaction for the transformation of eqn (3.13) is $\Delta H = \Delta q = -673 - (2 \times -327) = -19$ kilocalories per mole.

In general, the enthalpy change in a reaction can be calculated from

$$\Delta H^\circ_{reaction} = \Sigma \Delta H^\circ_{products} - \Sigma \Delta H^\circ_{reactants} \qquad (3.14)$$

$\Delta H^\circ_{products}$ and $\Delta H^\circ_{reactants}$ refer to the changes in enthalpy in forming the products and the reactants from their elements, $\Delta H^\circ_{reaction}$ being the required enthalpy change in the reaction. The summation signs indicate that all substances taking part in the reaction are to be taken into account. It is clear that the quantities involved must refer to specific amounts of substance and also, since H depends on temperature, pressure, etc., to specific conditions. This is done by always using what is called the enthalpy of the *standard state*. The standard state refers to one mole of the substance (or a one molal solution) at 25°C and one atmosphere pressure. The enthalpy of chemical elements in their standard state is always chosen to be zero. To indicate that standard states are referred to, a superscript zero is used, as in eqn (3.14). This equation therefore expresses the following idea: if a reaction is carried out in stages (by forming chemical compounds from their elements and then allowing them to react to give the desired products) then the algebraic sum of the enthalpy changes of the separate stages is equal to the enthalpy change of the reaction carried out in one step.

Example: Calculate $\Delta H^\circ_{reaction}$ for the hydrolysis of urea to carbon dioxide and ammonia,

$$H_2O + H_2NCONH_{2(aq.)} \rightarrow CO_{2(aq.)} + 2 NH_{3(aq.)}$$

This reaction, which takes place with the enzyme urease as a catalyst, is responsible for the smell which often permeates babies' clothes. The $\Delta H°$ of the reactants and products are shown in Table 3.1. The subscript (aq.) indicates a one molal aqueous solution.

TABLE 3.1

	$\Delta H°$(kilocalories per mole)
H_2O	−68·32
$CO_{2(aq.)}$	−98·69
$NH_{3(aq.)}$	−19·32
$H_2NCONH_{2(aq.)}$	−76·30

Writing out eqn (3.14) in full for the reaction we have

$$\Delta H°_{reaction} = \Delta H°_{CO_{2(aq.)}} + 2\Delta H°_{NH_{3(aq.)}} - (\Delta H°_{H_2O} + \Delta H°_{urea(aq.)})$$
$$= -98·69 - 2 \times 19·32 - (-68·32 - 76·30)$$
$$= 7·29 \text{ kilocalories.}$$

Notice that since the value of the overall enthalpy change is positive heat must be supplied to the system from the surroundings during the reaction.

Entropy

It is commonplace that as heat is added to water its temperature rises until it boils. When the water is boiling the temperature remains constant despite the addition of heat. What is happening to the energy that has been added to the water? The answer is that the energy has been used to create disorder among the water molecules. Water molecules which had been in the liquid have been boiled off and are now part of the vapour (steam) above the liquid. Whereas in the water the distance between neighbouring molecules was a more-or-less constant quantity, in the vapour this distance is a continually changing quantity, since the molecules move freely with virtually no interaction. There is thus less correlation between the positions of neighbouring molecules. There has been a similar change in the molecular velocities; previously rather similar to one another, they are now characterised by the Maxwell distribution and have a much larger spread. They too have become randomised. This change in the amount of ordering in a system is of much interest in thermodynamics, since it is brought about by means of energy changes. Hence a quantity known as *entropy* is introduced, which measures the amount of disorder or randomness in a system.

The symbol for entropy is S. We shall have a lot to say about

Thermodynamics

changes in the randomness of a system, and hence changes in S, but we shall never require to know the actual value of S as distinct from changes in this value. In fact the actual value of S can be defined and calculated (one takes the entropy of a so-called 'perfect' crystal at $0°K$, absolute zero, to be zero), but we do not need to make use of this. We shall define, not entropy itself, but change is entropy, as follows: the increase in entropy, ΔS, of a system due to absorbing an amount of heat Δq at temperature $T°K$ is given by

$$\Delta S = \Delta q/T \tag{3.15}$$

Example: Calculate the increase in entropy when one kilogram of ice at $0°C$ is converted into water at the same temperature. The latent heat of ice is 80 000 calories per kilogram, so $\Delta S = 80\,000/273 = 294$ calories per degree Kelvin per kilogram $= 5\cdot 3$ calories per degree kelvin per mole.

Using this concept of entropy we can proceed to formulate the second important principle of thermodynamics, known as —

The second law of thermodynamics

This law can be stated in many different ways. One of these is that the entropy of the universe can *only* increase (or stay constant) during any process. That is, any process which, if it occurred, would lead to a decrease of total entropy is not allowed. The second law is a statement about a 'direction' which is common to all spontaneous processes. As an example consider a tray that has upon it some black marbles and some white marbles arranged in rows of the same colour. There might be twenty rows each with twenty marbles. If the tray is now vigorously shaken the marbles will become disordered, that is, the entropy of the system will increase. If one persists in shaking the tray it is extremely unlikely that the marbles will reassemble in their original positions, though none of the laws of mechanics would forbid such a thing happening. Only the second law of thermodynamics, with its denial of such reordering (and consequent entropy decrease) being permissible, can be invoked to support our intuitive feeling that such an event could simply not happen.

The more marbles there are the more unlikely it is that shaking will restore order. Now, at the beginning of the chapter we said that thermodynamics makes statements about systems containing a large number of particles. Since the molar quantities of our everyday experience contain of the order of 10^{23} molecules the 'large number' involved is large indeed! Thus when such a real physical or chemical system is considered the probability of spontaneous reordering, once randomisation has occurred, becomes essentially zero. Reordering *can* be done but the second law tells us that a corresponding amount of disorder (entropy increase) will occur elsewhere in the universe, so that the latter's entropy either stays constant or increases. The growth of plants and animals is an obvious example of ordering, various chemicals being arranged to form complex organs. We must infer that the decrease in entropy due to the organism's growth is compensated by an equal or greater increase in entropy in the universe as a whole. In this particular instance the energy for the orderly growth comes eventually from the sun, whose energy output is attained by producing an increase in disorder amongst its nuclear fuels. The consequent increase in entropy of the universe is enormously greater than the niggardly decrease that terrestrial organisms can accomplish.

A process during which the entropy of the universe stays constant is known as a *reversible process*. In order to get some idea what type of process might be reversible, it is useful to consider the line of reasoning employed by the man who first formulated the second law of thermodynamics. He was a French military engineer named Carnot and he was interested in heat engines, particularly in the factors that governed their efficiency. Typically a heat engine, such as a steam engine, has a boiler at temperature $T_1^{\circ}K$ and a condenser at temperature $T_2^{\circ}K$, T_1 being greater than T_2. The details of the engine are not important except for the fact that it is *cyclic*. During each cycle a quantity of heat q_1 is taken from the boiler, an amount of work, W, is done and the remaining heat, q_2, is delivered up to the condenser. Since the engine is cyclic and returns to its initial state and initial internal energy (since the temperature returns to the initial value), the first law (eqn 3.1) shows that

$$W = q_1 - q_2 \qquad (3.16)$$

Carnot wanted to know whether another heat engine, perhaps running on a different substance than water, or of different design but working between the same two temperatures, could be more efficient. (We define *efficiency* as W/q_1, that is the work obtainable divided by the amount of heat taken from the boiler.) Clearly the engine should be mechanically frictionless in order to be as efficient as

Thermodynamics

possible. A characteristic of a completely frictionless system is that only an indefinitely small 'push' would be needed to set the system in motion. Also, to reverse the direction of motion only an indefinitely small push in the reverse direction would be required. Now Carnot found that for the maximum possible efficiency it is also necessary that the flow of heat in the engine be what we might call 'frictionless', that is such that an indefinitely small change in temperature could reverse the direction of heat flow. If this were so it would be possible to start the process, and then to stop it at some intermediate point and return to the original conditions leaving the surroundings completely unchanged. Such a process, with 'frictionless' heat flow, was called by Carnot a reversible process.

Clearly a reversible process is an idealization which cannot be attained in practice. For example, all temperature gradients would have to be infinitely small and hence the process would be indefinitely slow. However, Carnot showed that an engine that possessed both frictionless mechanics and frictionless heat flow (and which was therefore a reversible engine) had the maximum possible efficiency. Note that no mention has been made, nor is it necessary, of the details of such an engine. This is typical of thermodynamic arguments, whose generality we have already stressed.

Carnot showed that for *any* heat engine satisfying the above requirements

$$q_1/T_1 = q_2/T_2 \tag{3.17}$$

Thus the entropy change in a complete cycle of the engine is zero, since $\Delta S = q_1/T_1 - q_2/T_2 = 0$. Moreover, we can actually calculate a number for this best possible value of the efficiency. Elimination of q_2 between eqns (3.16) and (3.17) gives

$$W = q_1 \frac{T_1 - T_2}{T_1} \tag{3.18}$$

Hence the efficiency, W/q_1, is equal to $(T_1 - T_2)/T_1$. Thus only when the condenser is at a temperature of $0°K$, that is absolute zero, is it possible for *all* the heat taken from the boiler to be converted into work, with a consequent 100% efficiency. For all other condenser temperatures even a reversible engine is not perfectly efficient.

Example: If T_1, the boiler temperature is $450°K$ (superheated steam) and T_2, the condenser temperature is $300°K$ (cool water) then the efficiency of a perfectly reversible engine is

(450 − 300)/450 = 33%. In practice actual steam engine efficiencies measured by the ability to do work for a certain heat output from the boiler, are seldom more than 75% of this value.

Gibbs' free energy; its use in reaction analysis

In combining the results of the second law with the concept of enthalpy it is usual to define yet another quantity which characterises a system. The *Gibbs' free energy* of a system is denoted by G and defined as

$$G = H - TS \qquad (3.20)$$

We shall show that with this definition the change in Gibbs' free energy of a system during a reaction which takes place at a constant temperature and pressure is equal to the maximum work obtainable. The proof is as follows. If H changes by ΔH and S changes by ΔS, then the change in Gibbs' free energy, ΔG, is given by

$$\Delta G = \Delta H - T\Delta S \qquad (3.21)$$

if the temperature is constant, that is if $\Delta T = 0$. As we have already seen ΔH is equal to the amount of heat evolved or absorbed. Also $T\Delta S$ is equal to the amount of heat used in changing the entropy of the system (eqn 3.15). Thus eqn (3.21) states that at a constant temperature the maximum amount of work obtainable is lower than the energy evolved as heat by the amount used to alter the entropy. The energy converted into entropy has become *unavailable* energy, and if the entropy has increased this energy has gone to increasing the disorder in the system. The change in Gibbs' free energy, on the other hand, gives directly the amount of *available* energy that can be used to do work.

Example: Calculate the entropy change during the hydrolysis of adenosine triphosphate (ATP) given that $\Delta H = -4\cdot8$ kilocalories per mole and $\Delta G = -7\cdot4$ kilocalories per mole at a temperature of 36°C. Equation (3.21) connects ΔS with the quantities specified. Substitution of the values given yields $-7\cdot4 = -4\cdot8 - 309 \Delta S$. Hence $\Delta S = 2\cdot6 \times 10^3/309$ calories per degree kelvin per mole = $8\cdot4$ calories $°K^{-1}$ mole^{-1}. This entropy change is comparable with (in fact slightly more than) that occuring during the melting of ice.

If the change in Gibbs' free energy for a reaction is negative then the reaction will proceed spontaneously. This important statement is a direct consequence of the fact that entropy tends to increase.

Thermodynamics

In an isolated system an increase in entropy is equivalent to a negative ΔG since by 'isolated' we imply no change in enthalpy. If ΔG is zero then the system is in equilibrium and no change will occur. If ΔG is positive then the reaction will not occur spontaneously. It should be remembered though that a negative ΔG does not imply that the spontaneous reaction will occur *rapidly*; its rate might be undetectably low. However, as the reaction proceeds, no matter how slowly, G is decreased until it ultimately reaches a minimum value at which any change ΔG would be zero. The system has then reached equilibrium and the reaction stops. This is analogous to the mechanical example of a ball rolling down a slope into a hollow. The ball eventually comes to rest at the position of lowest potential energy.

Table 3.2 lists values of various thermodynamic variables for the water-ice transition at $-10°C$, $0°C$ and $+10°C$. It is interesting to note that the change in internal energy favours the water-ice transition at all three temperatures. Thus at $+10°C$ there would be a reduction of internal energy of 1529 cals mole^{-1} if the transition occurred. However, at the same temperature, the value of $T\Delta S$ is

TABLE 3.2
The water-ice transition

Temperature (degrees C)	ΔE (cals mole^{-1})	ΔH (cals mole^{-1})	ΔS (cals mole^{-1} °K^{-1})	$-T\Delta S$ (cals mole^{-1})	ΔG (cals mole^{-1})
−10	−1343	−1343	−4.9	1292	−51
0	−1436	−1436	−5.2	1436	0
+10	−1529	−1529	−5.6	1583	+54

such that ΔG for the transition is positive. Thus too much ordering is required and the transition does not occur spontaneously. At $0°C$ the value of $T\Delta S$ exactly balances ΔE giving a ΔG of zero for the transition, corresponding to an equilibrium situation. At $-10°C$ $T\Delta S$ has decreased still further and ΔG is negative, the transition being spontaneous at this temperature.

We will now relate ΔG directly to the position of equilibrium in a chemical reaction, making use of the *equilibrium constant* of eqn (2.21). At that point it was shown that a chemical reaction such as

$$A + B \rightleftharpoons AB$$

proceeds until concentrations of A, B and AB are reached such that

$$\frac{[A][B]}{[AB]} = K \tag{3.22}$$

where K, the equilibrium constant, depends on the temperature as well as on the nature of the reactants. Now as we have just seen, equilibrium is attained not only when the concentrations satisfy eqn (3.22) but also when the Gibbs' free energy of the system has reached a minimum value. Thus one expects there to be a relationship between Gibbs' free energy and the equilibrium constant. We shall state this relationship without proof; it is, that the change in Gibbs' free energy, when one mole of each reactant is converted into one mole of product under conditions such that each is maintained in its standard state, is given by

$$\Delta G^\circ = -RT \log_e(K) \tag{3.23}$$

As before the superscript zero denotes standard states.

Example: The equilibrium constant at 25°C for the reversible reaction

Glyceraldehyde 3-phosphate + dihydroxyacetone phosphate

$\rightleftharpoons$ fructose 1,6-diphosphate

is $8 \cdot 91 \times 10^{-5}$ moles. Calculate ΔG°. With $R = 1 \cdot 987$ cals mole^{-1} °K^{-1} and $T = 298$°K we substitute into eqn (3.23), finding $\Delta G^\circ = -1 \cdot 987 \times 298 \times \log_e (8 \cdot 91 \times 10^{-5}) = 1 \cdot 987 \times 298 \times 9 \cdot 33 = 5 \cdot 52$ kilocalories per mole.

Another area where Gibbs' free energy is used is in the calculation of the osmotic work done in transporting material across membrane boundaries. This type of transport is of the greatest importance to living organisms. All their metabolic processes take place in aqueous media and the organisms must obtain the raw material for these processes from the environment through their boundary membranes. Waste products are disposed of in the same way. The membranes are clearly capable of performing very complicated functions, permitting certain compounds to diffuse freely whilst forbidding the passage of other compounds which may be very similar in size or constitution. In addition to this type of differentiation, some membranes are capable of passing compounds *against* their concentration gradient, in direct contradiction of Fick's law. Thus not only must the structure of membranes be complex but in addition they must play an active role in the transport mechanism, rather than behaving as passive sieves. In spite of this

Thermodynamics

complexity and variability of the membranes, thermodynamics allows us to calculate the work done in any such osmotic transfer, given only the initial and final concentrations of the substance transferred, and ignoring the precise details of the transfer.

This connection between change in Gibbs' free energy and the initial and final concentrations in an osmotic process may be established by the following argument. We use the fact that solute concentration is related to the vapour pressure above a solution, and then, treating the vapour as an ideal gas, we find the work done as this vapour pressure changes during a concentration change.
The reasoning may seem a little tortuous but the fact that it works at all is a tribute to the power of thermodynamic arguments.

We start then with the knowledge that the presence of a solute in a solvent has the effect of lowering the vapour pressure — this is Raoult's law. So adding more solute to a solution will decrease the vapour pressure and the work necessary to bring about this pressure change can be calculated from eqn (3.4) if we assume ideal gas behaviour. This work is equal to the change in Gibbs' free energy. Thus, at constant temperature we have

$$\Delta G = \Delta W = -RT \log_e (V_f / V_i)$$
$$= -RT \log_e (P_i / P_f)$$

Now by Raoult's law P_i and P_f are proportional to C_i and C_f respectively, where these are the initial and final concentrations of solute. Hence $\Delta G = -RT \log_e (C_i / C_f)$, an equation which is more commonly written as

$$\Delta G = 2 \cdot 303 \, RT \log_{10} (C_f / C_i) \qquad (3.24)$$

where the logarithm is now to base ten for convenience of calculation and the minus sign has been taken inside the logarithm with the consequence that numerator and denominator are interchanged. Equation (3.24) enables the work done during the osmotic transfer of one mole of solute from an initial concentration C_i to a final concentration C_f at temperature $T°K$ to be found.

Example: Calculate the osmotic work done by the kidneys in secreting 0·158 moles of chlorine ions in a litre of urine at 37°C when the concentration of chlorine ions in the plasma is 0·104 moles per litre and in the urine is 0·158 moles per litre. Since 0·158 moles of chlorine ions are secreted, eqn (3.24) becomes (since as written it applies to one mole only)

$$\begin{aligned}\Delta G &= 0\cdot 158 \times 2\cdot 303 \times RT\ \log_{10}(C_f/C_i) \\ &= 0\cdot 158 \times 2\cdot 303 \times 1\cdot 987 \times 303 \times \log_{10}(0\cdot 158/0\cdot 104) \\ &= 40\cdot 7 \text{ calories} \\ &= 9\cdot 7 \text{ J}.\end{aligned}$$

A further example of the use of Gibbs' free energy will be found in the next chapter. There we will calculate the connection between the ΔG due to chemical reactions in an electric battery, and the amount of electrical work available from the battery. By this means a relation between the electric potential of the battery and the concentrations of its chemical components will be obtained.

4

ELECTROCHEMISTRY

Electrochemistry is the study of chemical reactions between polar (ionic) reactants, in particular of the manner in which such reactions are affected by electrical forces. These forces can be applied to a system either from outside, as with an electric battery, or internally by the presence of differences in concentration of ions of opposite polarity. Such concentration differences can arise only when the ions are free to move, that is when the substances concerned are dissolved in some suitable solvent. The solvent in all cases of practical interest is water, so we may redefine electrochemistry as the study of ionic reactions in aqueous solution. The subject is therefore of immense significance to biologists and biochemists, since virtually all vital processes do take place in such an environment. The interest of the biologist is heightened by the presence in living organisms of a variety of dividing membranes, most of them polar in nature and providing ideal boundaries across which electrical forces can act. Thus the propagation of electrical impulses down nerve fibres is one example of a topic whose study is only made possible using the results of electrochemistry.

The subject is usually approached using the techniques of thermodynamics to establish certain energy relations. We shall do this, but since some elementary knowledge of electricity and electrical laws is required we start with a brief statement of these.

The flow of electricity in metals

The current which can pass through a metal wire consists of a flow of negatively charged *electrons*. These electrons, in passing along the wire, travel down an electric *potential gradient* between two points at different potentials. These potentials are commonly provided by a battery with one positive and one negative terminal. The potential difference between the poles (terminals) when no current is passing is known as the *voltage* or *electromotive force*, and is measured in volts. Owing to a misunderstanding amongst early physicists about the nature of electricity, the convention universally used is that the electric current flows from the positive pole or *anode* to the negative pole or *cathode*. In fact of course the negative electrons flow from cathode to anode, but this discrepancy in direction is not usually troublesome.

Electric current is measured in *amperes*. One ampere is a flow of one coulomb per second where a coulomb is a certain quantity of electricity. The charge of an electron is $1 \cdot 6 \times 10^{-19}$ coulombs, so there are $1/(1 \cdot 6 \times 10^{-19}) = 0 \cdot 66 \times 10^{19}$ electrons in one coulomb. When a current flows down a potential gradient work is done. The electrical units are so defined that one ampere flowing through a potential difference of one volt does work at a rate of one watt.

Example: For how long must a 6V, 2W flashlight bulb be lit for one gram of electrons to flow through the filament? The mass of an electron is 9×10^{-31} kg so one gram contains about 10^{27} electrons. Since a power of 2W at 6V corresponds to a current of 1/3 of an ampere, that is to the flow of $1/3 \times (6 \cdot 6 \times 10^{18})$ electrons per second, the mass of electrons flowing per second is this number divided by 10^{27}, or about 2×10^{-9} g. So one must wait $1/(2 \times 10^{-9})$ seconds, that is 5×10^8 seconds or about 16 years, for one gram to flow.

Consider a metal wire with a potential difference of V volts between its ends. The current I amps which will flow is determined by the *resistance* of the wire. It is found experimentally that the current which flows is proportional to the voltage applied, and this is expressed algebraically as Ohm's law,

$$V = IR \qquad (4.1)$$

where R is the resistance of the wire. When V is expressed in volts and I in amperes, R is measured in ohms (Ω). Resistance is a quantity which depends both on the material through which the current is flowing and on the dimensions of the wire. For example, for a wire of uniform cross-section the resistance is proportional to the wire's length and inversely proportional to its cross-sectional area. Thus it is useful to introduce a quantity known as the *resistivity*, ρ, which is characteristic of the material and quite independent of its

Electrochemistry

dimensions. The resistivity is defined as

$$\rho = \frac{RA}{l} \qquad (4.2)$$

where R is the resistance of a wire of the material in question, A its cross-sectional area and l its length. The units of ρ are ohm cm.

Copper, for example, has a resistivity of $1 \cdot 72 \times 10^{-6}$ ohm cm. Thus the resistance of a 1 cm length of copper wire of uniform cross-section 10^{-4} cm² is $R = l\rho/A = (1 \cdot 72 \times 10^{-6} \times 100)/10^{-4} = 1 \cdot 72 \, \Omega$. It is often convenient to refer to the reciprocal of resistivity, known as *conductivity*, K.

$$K = 1/\rho = l/RA$$

K is measured in ohm⁻¹ cm⁻¹. We shall see shortly the reason for this preference. It relates to the behaviour of conductive solutions, to which we now turn.

Electrolytic conduction

It is well known that certain aqueous solutions can, like metals, conduct electricity. They are called electrolytic solutions and the solute is called an *electrolyte*. An electrolyte dissociates to some degree into positive and negative ions on entering solution. These ions carry a number of units of e, the charge on an electron, which is consistent with their chemical valency. For example, copper sulphate and hydrochloric acid dissociate in solution thus:

$$CuSO_4 \rightleftharpoons Cu^{++} + SO_4^{--}$$
$$HCl \rightleftharpoons H^+ + Cl^-.$$

In contrast to metallic conductors, the current in electrolytic solutions is carried by the ions. It is by no means obvious, though we shall assume it to be true, that Ohm's law is obeyed in such circumstances, nor is it clear how the conductivity is related to the nature and concentration of the various ionic species. This is the problem we now tackle.

Clearly the conductivity of an electrolyte solution is due to the ions experiencing a force when in an electric field and so travelling towards the appropriate electrode. Thus the conductivity is not only characteristic of the particular electrolyte but it also depends upon the ionic concentration. One may guess that, at least to the first approximation, conductivity will be proportional to the number of ions present, that is to the concentration. In order to remove this concentration dependence from the problem a quantity Λ called the *equivalent conductance* is defined as conductivity divided by concentration, that is

$$\Lambda = K/C \qquad (4.3)$$

where C is the concentration of electrolyte in equivalents per litre.

A brief word about chemical equivalence is in order here. The equivalent weight of a substance is defined as that weight associated with the transfer of 96 487 coulombs of charge in electrolysis. This amount of charge, known as the *Faraday*, is equal to Avogadro number multiplied by the charge on an electron, in coulombs. Thus the equivalent weight of a univalent ion like Cu^+ is equal to its atomic weight. The equivalent weight of divalent Cu^{++} is half the atomic weight of copper. This is because each ion carries two units of charge and hence only half the number of atoms in a gram-atom, that is one half of Avogadro's number, would need to be passed in electrolysis to carry 96 487 coulombs. Incidentally scientists often round off the Faraday to 96 500 coulombs per equivalent, only using the more exact figure when accuracy of that order is called for.

Example: For how long must a current of 5 mA be passed through a $CuSO_4$ solution in order to deposit one gram of copper at the electrodes? In such a solution the copper is divalent, so 96 487 coulombs will deposit 63·57/2 grams of copper. Thus one gram requires $2 \times 96\,487/63\cdot57$ coulombs, and since the current is 5 millicoulombs per second the time needed is $2 \times 96\,487/(63\cdot57 \times 5 \times 10^{-3}) \approx 6 \times 10$ seconds.

Equivalent conductance was introduced as a quantity characteristic of the particular electrolyte and, hopefully, independent of its concentration. Experimental work towards the end of the nineteenth century showed that, on the contrary, equivalent conductance still did depend to some extent on the concentration. Therefore its symbol Λ is usually written with a subscript c, Λ_c, to emphasise this dependence. There are a number of reasons for this dependence. For example, in the case of weak acids like acetic acid the change in Λ_c with concentration is due to the varying extent of ionic dissociation at different concentrations. On the other hand the same dependence was found even for the strong electrolytes such as NaCl and KCl, which dissociate completely in solution. For such substances the ions do not usually exist as independent entities in the solution, but tend to form clusters. For instance an ion of

Electrochemistry

Na⁺ might gather a group of three or four Cl⁻ ions about itself and the assembly would then move as a whole under the influence of an electric field. The details of such cluster formation depend on the concentration of the solution; the clusters are sometimes hydrated as well. The effect of cluster formation is to increase the friction or viscous drag associated with the movement of ions through the solution. Cluster formation is analogous to non-ideal behaviour in a gas, an effect which, it will be remembered, occurs when the molecular concentration is so great that the molecules do not have sufficient room to move entirely independently of one another.

For strong electrolytes the concentration dependence of equivalent conductance is found, experimentally, to be given by the equation

$$\Lambda_c = \Lambda_0 - a\sqrt{C} \qquad (4.4)$$

where C is the concentration and Λ_0 and a are constants. This relation between equivalent conductance and concentration is illustrated in Fig. 4.1. Since Λ_c plotted against $\sqrt{C}$ is a straight line it is easy to extrapolate back to zero concentration. The intercept is Λ_0, the equivalent conductance at zero concentration, and its significance is that it is the value of the equivalent conductance in the ideal, but experimentally unrealizable, situation that the ions exert no influence on one another. A similar approach was used in deducing the value of osmotic pressure at zero concentration in Chapter 2.

Since the ions are supposed to be totally non-interacting at this limit of zero concentration it is to be expected that each particular type of ion contributes a specific, and constant, amount to the overall equivalent conductance Λ_0. Thus the contribution of a Cl⁻ ion, say, should be the same whether its partner ions are Na⁺, K⁺, Cu⁺⁺ or any other cation.* This is borne out by the experimental results (see Table 4.1). Note that the differences in Λ_0 between NaCl and KCl, and between NaNO₃ and KNO₃, are equal. So are the differences between KCl and KNO₃, and between NaCl and NaNO₃. Thus it does appear that each ion contributes a definite amount to the equivalent conductance at zero concentration, and so it seems plausible that

---

* Cations are positive ions which are attracted towards the (negative) cathode. Anions are negative ions which are attracted towards the (positive) anode. The actual direction in which the current-carrying charges move is of prime importance in electrolytic conduction, as opposed to our general indifference to it in metallic conduction.

$$\Lambda_0 = \lambda_0^+ + \lambda_0^- \tag{4.5}$$

where λ_0^+ and λ_0^- are the contributions of the positive and negative ions respectively.

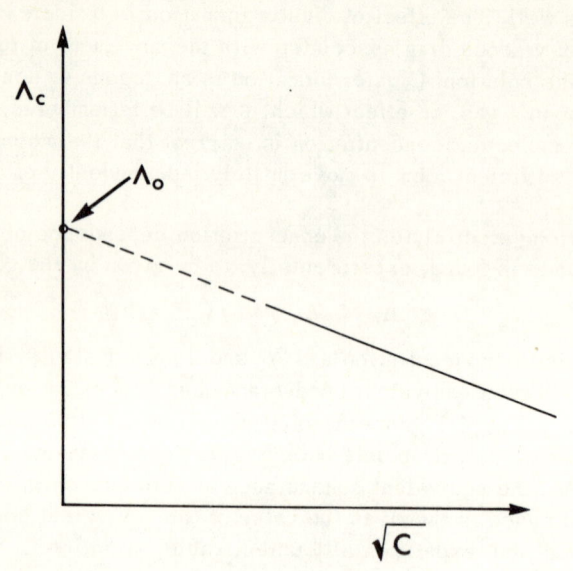

Fig. 4.1
A graph of the equivalent conductance, Λ_c, of a strong electrolyte plotted against the square root of its concentration $\sqrt{C}$.

TABLE 4.1

	Λ_0 (Ω^{-1}cm^2 equivalent^{-1} at 25°C)
NaCl	126·4
KCl	150·0
NaNO$_3$	121·5
KNO$_3$	145·1

Ionic velocity and transport number

We will now derive eqn (4.5) in a way which will show how the conductances can be rewritten in terms of the ionic velocities in the electrolyte. Consider a solution of a uni-univalent salt like NaCl, of concentration* C equivalents ml^{-1}. Suppose the solution

---
* Since the concept of a *normal* solution is still widespread we usually measure C in equivalents per litre. In this section only we use equivalents per ml and this gives rise to the factor of 1000 in eqn (4.7).

Electrochemistry

to be contained in a 1 cm cube and two opposite faces of the cube to be electrodes. Let the potential difference between these electrodes be X volts. Thus the ions are in an electric field of strength X volts per cm. Let the velocities of the positive and negative ions at this concentration be U_c and V_c cm s⁻¹ respectively. As will be realised these are only average velocities. The ions are accelerated by the electric field but also lose energy in collisions with other ions and solvent molecules. The situation is one of equilibrium, the large number of collisions resulting in some steady average velocity in spite of the continual accelerating force of the electric field.

The number of equivalents of positive ions per millilitre is C and so the number of coulombs of positive charge per millilitre is FC, where F is the Faraday. Similarly the number of coulombs of negative charge per millilitre is also FC. Now during the flow of current for one second the positive ions in a volume U_c and the negative ions in a volume V_c reach, and discharge at, their respective electrodes. Thus the amount of positive and negative charge passing through the electrodes is $U_c FC$ and $V_c FC$ coulombs per second respectively. The total current flowing is thus $(U_c FC + V_c FC)$ amps. Since the electrodes are of area one square cm, the *current density*, i (i.e. the current per unit area), is numerically equal to this total current, that is, $i = (U_c + V_c)FC$ amp cm⁻².

Now the average ionic velocity is proportional to the electric field strength; we state this as a reasonably self-evident fact, though the proof is not too easy. Therefore

$$U_c = U_c^0 X \text{ centimetres per second, and}$$
$$V_c = V_c^0 X \text{ centimetres per second}$$

where the quantities U_c^0 and V_c^0 are the velocities the ions would have in a field of one volt per cm, at this concentration. Consequently we may rewrite the current density as

$$i = (U_c^0 + V_c^0) FCX \text{ amp cm}^{-2} \qquad (4.6)$$

We now wish to relate the quantities in this equation to the equivalent conductance Λ_c. To do this we first find their relationship to the conductivity K. We know that $K = 1/\rho$, so, using eqn (4.2), $K = l/RA$. Using Ohm's law, which we assume to hold for the electrolyte, to substitute for R we get $K = lI/AV = (I/A)/(V/l) = i/X$ since I/A is i, the current density, and V/l is X, the potential gradient. Thus we have that

$$K = (U_c^0 + V_c^0) FC$$

from eqn (4.6) and hence, using eqn (4.3),

$$\Lambda_c = (U_c^0 + V_c^0)F/1000$$

Therefore in the limit that C tends to zero we have

$$\Lambda_0 = F(U_0^0 + V_0^0)/1000 \qquad (4.7)$$

where U_0^0 and V_0^0 are the velocities of the positive and negative ions in a field of one volt per cm, extrapolated to zero concentration. Equation (4.7) is of the same form as eqn (4.5). Comparing the two we see that

$$\lambda_0^+ = F U_0^0/1000$$
$$\lambda_0^- = F V_0^0/1000 \qquad (4.8)$$

This proves that eqn (4.5) is not only plausible but actually correct if we assign velocities U_0^0 and V_0^0 to the positive and negative ions respectively.*

From tabulations such as that of Table 4.1 it is possible to estimate values of λ_0 for various ions, and thus, by use of eqn (4.8), to calculate their ionic velocities (mobilities); a selection of these mobilities is given in Table 4.2.

TABLE 4.2

	Ionic mobility at 25°C (10^{-4} cm² s⁻¹ V⁻¹)
H^+	36·3
Li^+	4·0
K^+	7·6
Na^+	5·2
$1/2\,Ca^{++}$	6·2
OH^-	20·5
NO_3^-	7·4
Cl^-	7·8
$1/4\,Fe(CN)_6^{4-}$	11·5

Such estimates of U_0^0 and V_0^0 from experimental data give typical results of around 10^{-4} cm per second per volt per cm, which is a surprisingly small value. Both the hydrogen and the hydroxyl ions have relatively high mobilities greater than 10^{-3}. This is due to their motion involving a special mechanism of proton transfer through the water. We can also note that the velocity variation throughout the series of alkali metal ions Li^+, K^+, Na^+ is irregular,

---

* The quantities U_0^0 and V_0^0, referring to a potential gradient of 1 V cm⁻¹, are usually called ionic *mobilities*; they are just velocities measured under some standard conditions.

Electrochemistry

that is, it does not depend simply on the radius of the ion. This is due to hydration effects in cluster formation.

Since not all ions have the same velocity or even charge it is clear that in general the different ions are responsible for carrying different fractions of the current in electrolysis. For example, 70% of the current may be associated with the motion of positive ions and the remaining 30% with that of the negative ions. The electrolyte solution remains, of course, electrically neutral in all cases. (Although we say 'of course', the reader may wish to stop here and think how he would write out a proof of this last statement.) The *transport number* of an ion is defined as the fraction of the total current carried by that ion. Thus for univalent ions

$$t_c^+ = U_c/(U_c + V_c)$$
$$t_c^- = V_c/(U_c + V_c) \quad (4.9)$$

where t_c^+ and t_c^- are the transport numbers of the positive and negative ions respectively at a concentration C. Since all the current in the solution is carried by the ions, the sum of their transport numbers must be unity, that is

$$t_c^+ + t_c^- = 1 \quad (4.10)$$

Transport numbers can be calculated directly from values of ionic velocity. Thus, using Table 4.2 we find the transport numbers of Na⁺ and Cl⁻ in a solution of NaCl to be $5·2/(5·2 + 7·8)$ and $7·8/(5·2 + 7·8)$, that is 0·4 and 0·6 respectively. Since ionic velocities vary with the concentration, so also do the transport numbers. Some typical values at various concentrations are listed in Table 4.3.

TABLE 4.3
Transport numbers at 25°C

Concentration (gram-equivalents per litre)	HCl t^+	t^-	KCl t^+	t^-	NaCl t^+	t^-	KNO₃ t^+	t^-
0·00	0·821	0·179	0·491	0·509	0·394	0·606	-	-
0·01	0·825	0·175	0·490	0·510	0·392	0·608	0·502	0·498
0·10	0·831	0·169	0·490	0·510	0·385	0·615	0·502	0·498

Having discussed how electrolytes behave in solution, we turn to the effects that the electrodes themselves have on the situation. The simplest electrodes are metal rods dipped into the solution so we start with these. Such a combination of electrodes and electrolyte is called an electrochemical cell.

Electrochemical cells

If a metal rod, for example of copper or zinc, is dipped into water a steady difference of electric potential is set up across the metal-liquid interface. This is due to metal atoms floating off into solution as positive ions, the valence electrons remaining in the metal. That is to say, the rod literally dissolves in the water to some extent. Soon however a dynamic equilibrium is reached where the number of ions becoming reattached to the rod is balanced by the number leaving. If the rod is placed in a solution containing its own ions the same type of process will take place although the details will be different. For example, the potential difference set up across the metal-liquid interface will now depend upon the concentration of the solution. Thus a zinc rod placed in a one molar zinc sulphate solution becomes negatively charged with respect to the solution, that is, positive zinc ions have a greater tendency to leave the rod and enter solution than those ions already in solution have to return to the rod. On the other hand a copper rod becomes positively charged with respect to a one molar copper sulphate solution. This results in a distribution of charges across the metal-liquid interface, termed an electrical 'double layer'. It arises because the valence electrons left, say, in the metal congregate at the surface thus attracting a layer of positively charged ions. There thus exists a potential difference across the interface which is referred to as the *reversible electrode potential*. This potential* is concentration dependent since altering the electrolyte concentration changes the degree to which metal ions will dissolve in the solution. Strictly speaking, it should be said that the potential's value at a given temperature depends on the *activity* of the solution rather than on its concentration. Activity is a concept which was introduced because certain quantities, like osmotic pressure, did not vary with concentration in the manner predicted by the ideal gas analogy.

---

* The potential is called *reversible* because under certain conditions the reactions that occur in electrochemical cells can be made to satisfy the conditions for reversibility outlined in the previous chapter. We shall see the significance of this shortly.

Electrochemistry

These effects of non-ideal behaviour are now fairly well understood and a book on physical chemistry (such as that by Ladd and Lee, p. 220) should be consulted by the interested reader. At low concentrations the distinction between activity and concentration becomes less important, and the difference between the two will be ignored in this book.

A single metal rod dipping into a solution is sometimes called a *half-cell*, and an interesting problem arises if the measurement of its electrode potential is contemplated. The passage of current would cause the potential difference to change a little, so it is necessary to measure the potential on 'open circuit', that is, drawing no current, using an instrument such as a potentiometer or a valve voltmeter. But in order to complete the circuit a *second* metal electrode would have to be introduced into the solution and this would introduce another (different) electrode potential. The measurement would record the difference of the two potential differences, not the potential difference of interest. There is no such problem in measuring the potential difference between the electrodes of a cell consisting of two half-cells since the valve voltmeter (say) can be connected to the two electrodes directly.

To avoid the difficulty with half-cells it is conventional for all reversible electrode potentials to be determined by forming a cell out of the half-cell in question and a standard reference half-cell. The difference in potential between the electrodes of this cell is then measured with a valve voltmeter or potentiometer, on open circuit. The result is often called the *electro-motive force* (e.m.f.). This method is tantamount to defining the reversible electrode potential of the reference half-cell to be zero and measuring all potentials with respect to it. The reference chosen is the standard hydrogen electrode. This consists of hydrogen gas at unit partial pressure in contact with a solution of hydrogen ions of unit activity. The hydrogen is bubbled over the surface of a platinum electrode with the temperature fixed at 25°C. Since there is always a danger of explosion when hydrogen is used this electrode is often replaced by a more convenient subsidiary standard, for instance a calomel reference electrode. For further information on such cells, see the bibliography.

We will now relate these reversible electrode potentials to the ion concentration of the electrolyte solution and to the type of reaction occurring in the cell. To deal with the latter point first, we will consider as an example the cell shown in Fig. 4.2. Such

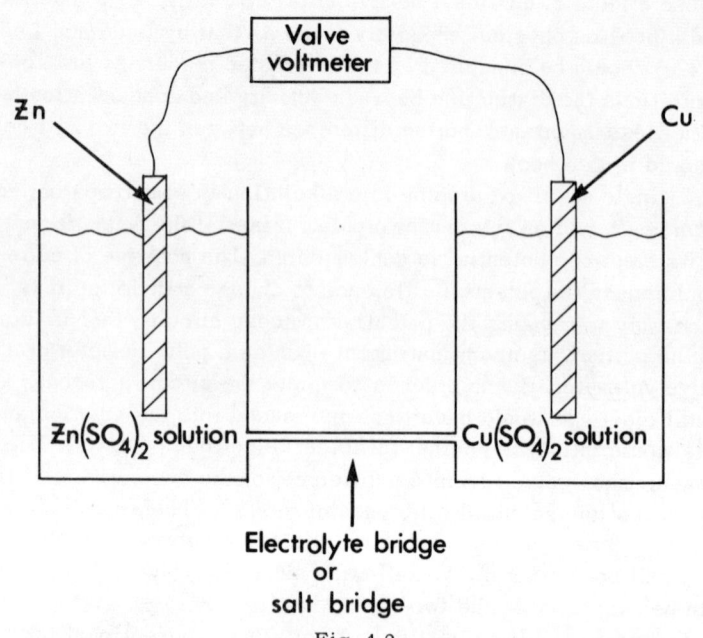

Fig. 4.2
A Daniell cell, consisting of zinc and copper electrodes in their respective sulphate solutions, the two half-cells connected by a tube of electrolyte.

a cell, with zinc and copper electrodes in their respective sulphate solutions, is known as a Daniell cell. As we draw it we see it consists of two half-cells connected by a narrow electrolyte bridge. If both solutions are one molar then the zinc rod will be negatively charged and the copper rod positively charged. If they are connected via an electrical resistance then a current of electrons will flow from the zinc to the copper. As this occurs more zinc atoms will float off into solution, each leaving behind two valence electrons, in an attempt to maintain the reversible electrode potential. Similar reactions will occur in order to keep the copper rod positively charged. The processes are

$$Zn \rightarrow Zn^{++} + 2e^-$$
$$Cu^{++} + 2e^- \rightarrow Cu$$

Therefore the overall cell reaction is

$$Zn + Cu^{++} \rightarrow Zn^{++} + Cu$$

Electrochemistry

The reactions in the individual half-cells are either reducing or oxidising reactions. For example, in the reaction $Zn \rightarrow Zn^{++} + 2e^-$ the zinc atom is initially in the reduced state since neutral zinc has more electrons than the Zn^{++} ion. Therefore this particular reaction is one of oxidation. Half-cells also exist in which both the oxidised and the reduced forms are present as ions in solution. An example is a solution of ferrous and ferric ions with a platinum electrode. Electrons are supplied or accepted by the platinum so that the appropriate oxidation or reduction reaction can proceed. Whatever the details of the half cell, we may conclude that all reactions result in the transfer of one or more electrons.

We are now in a position to relate the electrode potentials directly to the ion concentrations. Thermodynamic arguments are used. The electric current drawn from a cell can do work, for instance by lighting a flashlamp bulb. The energy is provided by the chemical reaction taking place in the cell. Thus the change in Gibbs' free energy in this reaction can be equated to the amount of electrical work done. Now we have seen that the definition of Gibbs' free energy requires that the conditions be restricted to those of constant temperature and pressure, and reversibility of the reaction. Under such conditions (which hold not only for electrochemical cells in the laboratory but also for all the varied types of electrochemical reactions in living organisms) we may equate ΔG with the electrical work done. This procedure leads to an equation for the reversible electrode potential, which we shall call $\mathcal{E}$, in terms of the ion concentration C of the electrolyte solution. We shall state this equation without proof. It is

$$\mathcal{E} = \mathcal{E}_0 + \frac{RT}{n\mathcal{F}} \log_e C \qquad (4.11)$$

where $\mathcal{E}_0$ is a constant, characteristic of the cell reaction and called the *standard electrode potential*, n is the number of electrons transferred in the cell reaction (e.g. two, in the reaction $Zn \rightarrow Zn^{++} + 2e^-$) and $\mathcal{F}$ is equal to the amount of work obtained when Avogadro's number of electrons fall through a potential difference of one volt. It is conventional to evaluate the constants in eqn (4.11), at a temperature of 25°C and using logarithms to base 10. When this is done, eqn (4.11) is rewritten as

$$\mathcal{E} = \mathcal{E}_0 + \frac{0 \cdot 059 \log_{10} C}{n} \qquad (4.12)$$

Table 4.4 lists the standard electrode potentials, $\mathcal{E}_0$, for a variety of half-cell reactions.

To illustrate the use of eqn (4.12) consider a Daniell cell in which the concentration of the two solutions is the same, C say, and n has the value two. The reversible electrode potentials, $\mathcal{E}_{Cu}$ and $\mathcal{E}_{Zn}$, of the two half-cells are given as

$$\mathcal{E}_{Cu} = (\mathcal{E}_0)_{Cu} + \frac{0.059}{2} \log_{10} C$$

$$\mathcal{E}_{Zn} = (\mathcal{E}_0)_{Zn} + \frac{0.059}{2} \log_{10} C$$

Provided that no extra potentials are introduced by the salt bridge, the potential E of the cell on open circuit is given by the difference between these two electrode potentials, that is

$$E = (\mathcal{E}_0)_{Zn} - (\mathcal{E}_0)_{Cu}$$
$$= 0.763 - (-0.337)$$
$$= + 1.1 \text{ volts}$$

where the numerical values are taken from Table 4.4.

If the concentrations had differed, that is, if the concentration of the solutions had been C_{Cu} and C_{Zn} respectively, then the cell potential on open circuit would have been given by

$$E = (\mathcal{E}_0)_{Zn} - (\mathcal{E}_0)_{Cu} + \frac{0.059}{2} \log_{10} \left(\frac{C_{Zn}}{C_{Cu}}\right) \quad (4.13)$$

If, for example, C_{Cu} were 0.1 N and C_{Zn} were 1.0 N, then we would have

$$E = 0.763 - (-0.337) + \frac{0.059}{2} \log_{10}(10)$$
$$= 1.13 \text{ volts}$$

Another use of eqn (4.12) is to calculate $\mathcal{E}$, the reversible electrode potential, of a given combination of electrolyte and electrode. If the concentration of the electrolyte is measured and its value substituted in eqn (4.12), together with a value for $\mathcal{E}_0$ taken from the table, then $\mathcal{E}$ is obtained. This would be the e.m.f. of a cell formed by combining this electrode and electrolyte with a standard half-cell. On the other hand, the reverse procedure could be used,

Electrochemistry

TABLE 4.4

Standard electrode potential $\mathcal{E}_0$ for some half-cell reactions at 25°C

Half-cell reaction	$\mathcal{E}_0$ (volts)
Ba → Ba^{++} + 2e$^-$	+ 2·90
Al → Al^{+++} + 3e$^-$	+ 1·66
Zn → Zn^{++} + 2e$^-$	+ 0·763
Pb → Pb^{++} + 2e$^-$	+ 0·126
H$_2$ → 2H$^+$ + 2e$^-$	0·000
Cu → Cu^{++} + 2e$^-$	− 0·337
Cu → Cu$^+$ + e$^-$	− 0·521
Fe^{++} → Fe^{+++} + e$^-$	− 0·771
2Cl$^-$ → Cl$_2$ + 2e$^-$	− 1·360
2F$^-$ → F$_2$ + 2e$^-$	− 2·87

and this is more common in practice. Measurement of the e.m.f. allows one to calculate electrolyte concentrations by use of eqn (4.12). Such a voltage measurement is one way in which the concentration of the electrolyte in automobile batteries can be measured, though it is often more convenient to measure the density of the solution using a hydrometer.

Concentration cells and *pH* values

A cell can be made from two half-cells which have the same electrodes and electrolyte and differ only in the concentration of the solution. Consider the cell shown in Fig. 4.3(a). If the concentrations of the two silver nitrate solutions are different then there will be different electrode potentials for the two half-cells, the overall cell potential being equal to the difference of these. Such a cell is called a *concentration cell*. In order to calculate the potential of such a cell, eqn (4.12) is applied to each half-cell. The difference is then taken and, the standard electrode potentials being equal, the result is

$$E = \frac{0 \cdot 059}{n} \log_{10}\left(\frac{C_1}{C_2}\right) \quad (4.14)$$

In the case of silver nitrate *n* is one, so a factor of ten difference in concentration would give rise to a cell potential on open circuit of 59 mV. Notice that there is no resultant chemical reaction in such a cell. The electrical energy is produced entirely by changes in concentration of the solution, the weaker becoming more concentrated and vice versa. The accompanying changes in Gibbs' free energy are given by eqn (3.24).

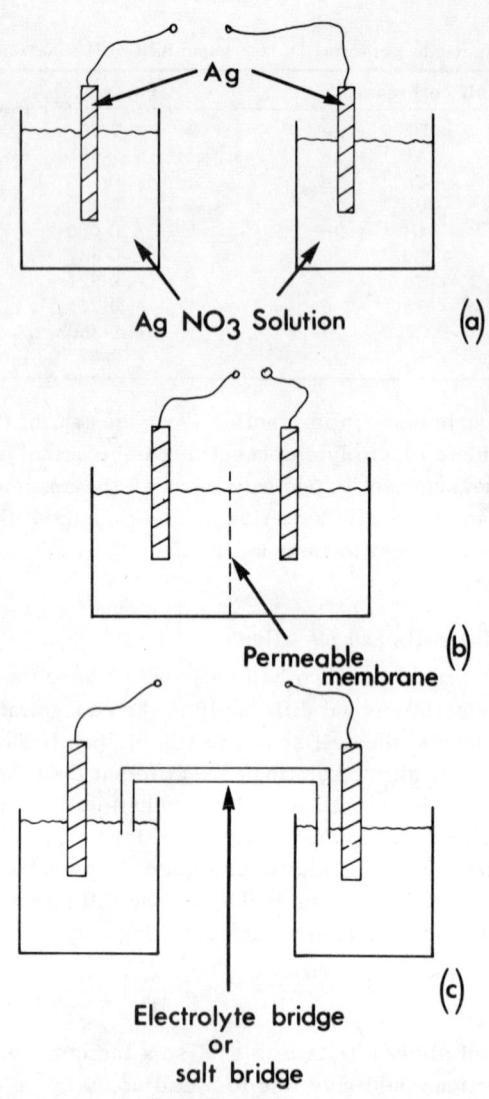

Fig. 4.3
A silver nitrate concentration cell, showing three different ways of connecting together the solutions of the two half-cells. (a) No connection at all. (b) Connection via a permeable membrane. (c) Connection via an electrolytic bridge.

Electrochemistry

If the two halves of a concentration cell are allowed to come into contact via a permeable membrane (Fig. 4.3b), it is found that the resultant cell potential differs from the case when there is no possibility of ion transport between the half-cells, as in Fig. 4.3(a). This is due to another potential difference being set up *across the membrane*. This potential is known as a *liquid junction potential* or a *diffusion potential*, and it arises as follows. Since the ion concentration differs on either side of the membrane there will be a nett diffusion of ions towards the side of low concentration. In general ions of one charge will move faster than those of the opposite charge. This will cause a charge difference across the membrane which will slow down, and eventually bring to a halt, the nett diffusion of ions. In this condition the layers of charge at the boundary result in a difference of potential across it. Thus the potential of the cell of Fig. 4.3(b) is less than that of Fig. 4.3(a) by the value of the liquid junction potential.

Such potentials can upset attempts at accurate measurements, and the method of eliminating them is to arrange two of them of opposite sign so that they cancel each other out. It can be shown that if the half-cells are joined by a tube containing a solution of ions of equal transport number, then the liquid junction potentials set up at each end of the tube will cancel. Such a connecting tube is called a *salt bridge* or *electrolytic bridge*, and suitable electrolytes can be found from inspection of Table 4.3. A solution of potassium chloride or ammonium nitrate, formed into a gel with agar to prevent physical movement of the electrolyte, is commonly used. Though bulk flow of the liquids is thereby restricted, ions can still travel from one half-cell to the other (see Fig. 4.3c).

Concentration cells have been discussed primarily because of their importance in biological systems. Differences of electric potential across cell membranes arise from differences in ion concentration. The potential differences can be set up very rapidly, say in the order of milliseconds, by the rapid diffusion of ions such as Na^+ and K^+. Such changes are responsible, for example, for the rapid propagation of electrical impulses down nerve axons. Although there are actually no electrodes involved the potential differences may still be calculated using eqn (4.14). Although measurement of the concentration differences involved is naturally rather difficult, it has been done; in particular many such studies have been carried out on the common squid, which has some exceptionally large nerve axons. It is found that the KCl concentration can differ by a factor as large as 19 across the enclosing membrane, giving rise to a potential difference of $E = 0\cdot059 \times \log_{10}(19) = 75$ mV. Measurements of these potentials and of their rates of change are an important part of modern physiology, and a reference is cited on p. 220.

It is possible to measure the concentration of hydrogen ions by forming a concentration cell from a standard half-cell and a hydrogen electrode immersed in the solution under study. The hydrogen ion concentration, C_H, is an important parameter in many biochemical situations, and because the range of values which it can assume is enormously large (a factor of 10^{14} is needed to span the range) one usually quotes not C_H itself but a subsidiary quantity known as *pH*, defined as the logarithm of the reciprocal of C_H, that is

$$pH = \log_{10}(1/C_H)$$
$$= -\log_{10}(C_H) \quad (4.15)$$

C_H is here the concentration in gram-ions per litre. The *pH* scale varies from 14 for a strongly alkaline solution, through 7 for a neutral solution, or pure water, down to 0 for a strongly acid solution. Thus the *pH* of a $0\cdot01$ N aqueous solution of hydrochloric acid is 2 and a sodium hydroxide solution of the same normality has a *pH* of 12. One can calculate these values by remembering that a normal solution of a completely dissociated acid or alkali has a *pH* of 0 or 14 respectively; the *pH* of a given solution differs from these values by the number of powers of ten that its normality differs from unity.

To revert to hydrogen concentration cells, the potential difference at 25°C between two hydrogen electrodes placed in solutions of hydrogen ion concentration C_H and C'_H is given by eqn (4.14) with n set equal to one.

$$E = 0\cdot059 \log_{10}(C_H/C'_H) \quad (4.16)$$

If the potential is developed between a standard hydrogen electrode, for which C'_H equals one, and a hydrogen electrode immersed in the solution of unknown *pH*, then

$$E = 0\cdot059 \log_{10}(C_H)$$
$$= -0\cdot059\,(pH)\text{ volts} \quad (4.17)$$

If the temperature is different from 25°C, say T, we have the more general equation

$$E = -0\cdot001984\,T\,(pH) \quad (4.18)$$

Electrochemistry

Many commercial instruments to measure *pH* values are based on this principle of finding the e.m.f. of a hydrogen concentration cell at a fixed temperature. For reasons of convenience electrodes other than hydrogen are usually used.

Example: Suppose the measured electrode potential of a hydrogen electrode at 25°C is 0·708 volts, then from eqn (4.17) the *pH* value of the solution under test is 0·708/0·059 = 12. Thus the hydrogen ion concentration is 10^{-12} gram-ions per litre and the solution is strongly alkaline.

We have already noted that in many biological systems the number of hydrogen ions present, that is the *pH* value, is an important parameter. We mention in passing that there are also situations in which the *total amount* of acid present, that is the amount of titrable acid, is the quantity of more direct relevance. The two quantities are not necessarily equivalent because of the very low degree of dissociation of some weak acids.

5

OPTICS AND THE WAVE NATURE OF LIGHT

The topics we have discussed so far have needed only a small amount of theoretical understanding in order to derive some quite complicated results. For the remainder of the book this aim of ours will become a little more difficult to realise, since the subjects we treat are either relatively new in themselves, and hence only recently understood even by physicists — this is the case in Chapters 6 and 7; or, as in our treatment of optics here, the subject-matter is basically rather mathematical. On the other hand, the end results are of such everyday importance that we think it worthwhile to discuss the fundamentals qualitatively, and then to say 'it can be shown that...' to justify some particular numerical result.

A knowledge of the nature of light is essential before its characteristics and applications can be understood. We therefore start with a brief review of theories about light, and then show that if we make some simple approximations and assume light rays travel in straight lines we can derive the whole theory of geometrical optics. We then discuss three important characteristics resulting from light actually being a wave motion; these are its interference, its diffraction and its polarisation. Finally we mention lasers as an example of a field where many rapid advances are being made. For further reading on the topics discussed in this chapter see the books by van Heel and Velzel and by Weale, p. 220.

The nature of light

The scientific study of optics was virtually begun by Newton, who thought that beams of light consisted of streams of tiny particles or 'corpuscles'. This was a very natural assumption since light travels in straight lines (at least so far as Newton could tell) and it was difficult to conceive of anything other than a material object doing so. Newton's influence was such that his corpuscular theory was accepted almost universally, until experiments by Young* and others in the early nineteenth century demonstrated clearly that light is a *wave motion*. They showed that light rays could interfere with one another, that they could bend slightly round sharp edges, and that they could have various directions of vibration — all these are characteristics that only waves have.

Most waves that we are familiar with are vibrations of some material object — sea waves are a manifestation of the varying height of the water, sound waves of the alternate rarefactions and compressions of the medium's density, and so on. Light appears to have no such material medium associated with it. The clearest evidence for this is that we can see the sun, stars and planets through the vacuum of space — their light must travel through the vacuum. Light in fact is one of a class of waves called *electromagnetic*, being associated with oscillations in the magnitude of electric and magnetic fields. Such fields can exist in a vacuum and so light, along with all other electromagnetic waves such as radio waves and X-rays, can travel in a vacuum. Experiments show that these waves actually do exert a force on electrons in their path, thus proving that electric fields are involved. This indeed is how our radio receivers work, the electrons in the aerial being jostled up and down by the electric field of the radio waves. Although a magnetic field is also present, its effect is usually utterly negligible compared with that of the electric field and we shall ignore it.

We now summarise the main properties of all these electromagnetic waves. First, they all travel with exactly the same velocity in a vacuum (and, to a very good approximation, in air also). This velocity is universally known as c, the 'velocity of light'. Its value is very close to 3×10^8 m s^{-1}. This finite, though large, velocity accounts for the well-known time lag in radio communications with

---

* Thomas Young was an extraordinary man; not only did he establish the wave theory of light, but he did much valuable work on the principles of insurance, the dynamics of blood flow, and Egyptology, coming very close to deciphering the hieroglyphs on the Rosetta stone. He also had very prominent eyes. This enabled him to test various theories then in vogue as to how the eye accomodates for near vision. For instance he disproved a suggestion that the eyeball actually elongates by rotating his eye inward and clamping iron rings, one in front of and one behind the eyeball. The pressure he felt did not change during accomodation for near vision, showing the eye to remain the same shape.

Optics and the Wave Nature of Light

astronauts. The wavelength λ is related to the frequency ν by eqn (1.42), $c = \lambda \nu$. The frequency of visible light is between 10^{14} and 10^{15} Hz (the hertz, Hz, meaning one cycle per second, is the standard unit of frequency) and the wavelengths range from about 400 to 700 nm. On entering a medium other than vacuum the frequency remains constant but the wavelength and velocity alter. In fact the velocity v is always *less* than the velocity in a vacuum, c. The ratio of the two velocities, $n = c/v$, is known as the refractive index of the material. For most transparent liquids and solids n lies between 1·2 and 1·8, being 1·33 for water, 1·5–1·6 for various kinds of glass, and 1·42 for the lens of the human eye.

Light is an example of what is called 'transverse' wave motion, the oscillating electric and magnetic fields being perpendicular both to each other and to the direction of motion (see Fig. 5.1). A wave as shown in this figure is said to be *plane polarised*, the electric field being in a definite plane which in this case is the plane of the paper. This gives certain directional properties when it interacts with matter. Thus an electromagnetic wave is often referred to as a *vector* wave motion.

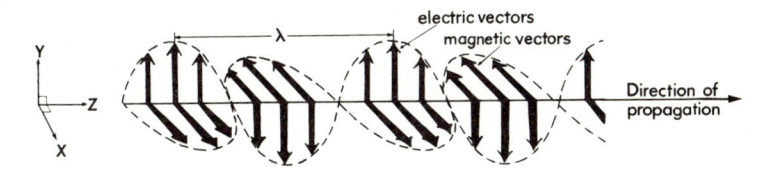

Fig. 5.1
Plane polarised monochromatic light. The electric field vectors are in the Y-Z plane and the magnetic field vectors in the X-Z plane.

These then comprise the main properties of light as it travels through space. We must now say a brief word about how light is emitted. A very hot object such as the sun or the filament of an electric light bulb contains atoms and electrons in violent thermal motion, and the agitation of these electrical charges is accompanied by the emission of light. This light can be of any wavelength, dependent solely upon the vibrational frequency; also its plane of polarisation lies at random directions. Moreover the light is emitted not as a continuous, infinitely-long wave but as a series of short *wave trains*, each corresponding to a particular vibrational frequency.

Because of the large number of collisions of the atoms and electrons in the incandescent material the phase and polarisation direction of the light changes abruptly at short intervals, these changes corresponding to the end of one wave train and the beginning of the next.

Almost the same remarks apply to glowing vapours such as fluorescent lights and neon tubes. The only difference is that the light is emitted by excited atoms as they decay to their ground states and lose energy. As we shall see in Chapter 6, only certain fixed amounts of energy are allowed to leave such an atom and thus the wave trains are of a definite rather than a variable wavelength. By using colour filters to remove unwanted wavelengths such light can be rendered of a unique wavelength, that is *monochromatic*. Finally it is useful to remark that a single wave train from most natural sources lasts about 10^{-9} seconds only. In this time light travels a distance of about $3 \times 10^8 \times 10^{-9}$ metres = 0·3 m, that is about one foot, so each wave train is about one foot long. Later on we shall sometimes call these wave trains *photons*.

Although light is a wave motion, for many purposes it is possible to ignore this fact and treat light as a ray travelling in a perfectly straight line. Such an approximate theory is called *geometrical optics*. It works in practice because the wavelength of light is so short. In later sections we shall see its deficiencies, but now we describe some of its well-known results and how they may be arrived at.

Geometrical optics: the design of optical instruments

Geometrical optics is the study of how optical instruments produce images by passing light rays through a series of lenses, prisms and mirrors. The reader will probably be familiar with the lens laws and with the laws for constructing image positions by drawing rays through a thin lens. If not it would be useful to consult an elementary textbook (see the bibliography) in order to become acquainted with image formation in, say, the compound microscope.

The first aim of this section is to show that the whole of geometrical optics stems from a single statement known as *Fermat's principle*. The other aim is to outline a modern approach to optical calculation in instrument design, a once sterile topic which has been revivified by the advent of modern computers.

---
* One of the intriguing features of physics is the manner in which a way of thinking in one branch so frequently proves useful in others. Fermat's principle is a case in point — it was originally proposed in mechanics but has proved itself equally useful in optics.

Optics and the Wave Nature of Light

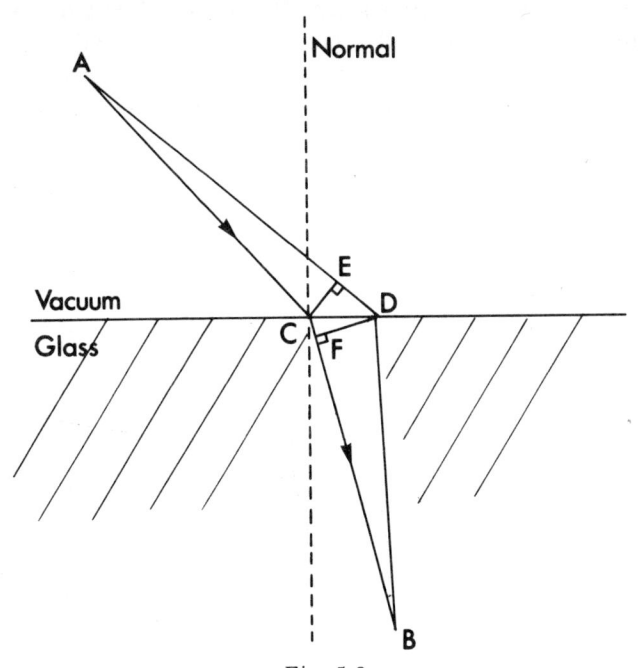

Fig. 5.2
ACB is the actual path of the light rays. *ADB* is a slightly displaced path. *CE* is perpendicular to *AD* and *DF* is perpendicular to *CB*.

To derive all the results of geometrical optics it is necessary to know only a handful of facts; that light rays travel in straight lines through homogeneous media, and that they obey the laws of reflection and Snell's law describing how the rays bend on going from one medium to another. All these results are easily derived from Fermat's principle which states that the time taken by a light ray to travel from one point to another is *the minimum that it can possibly be*. We can illustrate what this means by reference to Fig. 5.2 where we show two of the possible paths by which light could travel from point *A* to point *B*. Suppose that the actual path taken by a light ray is via point *C*. Let us now choose another path which differs from the actual one by some small amount which we call δ. Fermat's principle states that the time taken to travel along this slightly different path will be greater than the travel time along the true path, using in each case the speed of light appropriate to the medium the light is in. Thus on a graph (Fig. 5.3) of time taken versus δ the point Y, corresponding to the actual path of the ray, will lie lower than the point X corresponding to the slightly different path. The point Y is in fact lower than any other possible point. More precisely, therefore, Fermat's principle is that the time taken by a light ray is *stationary*, i.e. that the curve of Fig. 5.3 is virtually flat in a small region either side of Y. This means that there is effectively no change in travel time for a slight perturbation of the ray's path from the actual path.

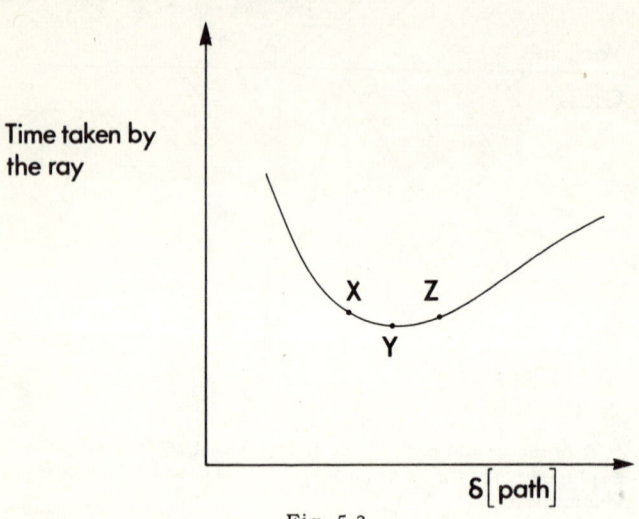

Fig. 5.3
The actual path taken by the ray corresponds to the minimum travelling time, that is to point Y on the graph.

As thus stated Fermat's principle may seem rather abstract. However we shall now apply it to a simple example to derive Snell's law of refraction.

Consider a light ray starting at point A in vacuum and reaching point B in glass (see Fig. 5.2). Clearly the light will not travel along the straight line AB because the journey time can be reduced by travelling somewhat further in vacuum where the light velocity is greater than in glass (by a factor of about 1·5). Thus a path such as ADB is preferred. The problem is to decide the position where the ray intersects the glass.

Suppose route ACB was the actual path taken by the light ray and ADB was a near-by path. The difference between ACB and ADB is an example of what is meant by δ i.e. a small perturbation or variation in path. Fermat's principle tells us that the travel times

Optics and the Wave Nature of Light

are the same for *ACB* and *ADB*, if *D* and *C* are very close. Thus the extra time spent travelling in vacuum along *ED* (when using route *ADB*) is exactly compensated for by the saving in the glass due to missing section *CF* (when using route *ACB*)

$$\therefore \frac{ED}{c} = \frac{FC}{v}$$

where c and v are the velocity of light in vacuum and glass respectively.

$$\therefore \frac{CD \sin i}{c} = \frac{CD \sin r}{v}$$

where *i* and *r* are the angles of incidence and refraction respectively (angles between appropriate ray and the normal).

$$\therefore \sin i = n \sin r \qquad (5.1)$$

where $n = \frac{c}{v}$ is the refractive index of glass. This is the familiar result that the ratio of the sines of the angles of incidence and refraction is equal to the ratio of the velocities in the two media.

One simple example of eqn (5.1) will suffice. The refractive index of water is 1·33, which means that when light rays enter water they get bent towards the normal to the surface according to eqn (5.1) with *n* set equal to 1·33. Thus rays coming almost horizontally along the surface, with *i* roughly equal to 90°, will, as they pass through the surface, be bent and enter at an angle given by $\sin r = 1/1·33$, that is $r = 48·8°$. The reader can draw a sketch to convince himself that, as a result, a fish's view of the world above the surface is restricted to a cone of vertex angle $2r = 97·6°$.

Among the other results which follow from Fermat's principle are the laws of reflection (angle of incidence equals angle of reflection), the straight paths of light rays in homogeneous media, and the fact that the travel time is the same for all rays leaving an object point and reaching a common image point. As an exercise, try to convince yourself of the first of these results.

The main bulk of geometrical optics is the derivation and use of laws governing the path of light rays through optical systems. The usual approximation both for thin and thick lens is to assume that all rays are close to the axis (paraxial approximation). Under this restriction all curved surfaces can be assumed to be sections of spheres. Thus an ellipsoid would be approximated by a sphere whose radius of curvature was equal to that of the ellipsoid at the axis. This allows one to derive simple expressions for magnification in terms of object and image distances. The use of these results gives excellent insight into the essential features of optical instruments. In practice, of course, the light rays are not restricted to a region close to the axis. Clearly the smaller the radius of curvature of the surface, the closer to the axis one has to get for these laws of geometrical optics to be applied. Small radii of curvature turn out to be essential for achieving high magnifications and hence are commonly encountered. In practice, then, the images formed by lenses are not perfect. The lens is said to have aberrations. Typically the axial and outer regions of the lens have different imaging properties thus the actual image is not simple at all.

Let us consider exactly what is involved in designing a particular optical system, e.g. a camera lens or microscope objective. The only piece of 'physics' involved is eqn (5.1), Snell's law. The problem is difficult because there will be many surfaces of differing shape, materials of various refractive indices and certain distances separating components. All of these must be varied until the best possible image-forming properties are achieved. In general this does not mean trying to remove aberrations. The normal approach is to attempt to get the error introduced by one type of aberration cancelled by an opposite error due to a different kind of aberration. The problem is difficult and is not made easier by trying to derive equations describing the system which can be solved in order to find the best optical design.

The classic method of design and the one still taught in schools is ray-tracing, a process whereby individual rays are traced through the system. The change of direction at each surface is calculated from Snell's law. In this way the position at which each ray intersects the desired image plane can be determined.

This approach has been given a new lease of life by the advent of fast modern digital computers. Although the amount of computation involved in finding the new direction of a ray on entering a different material is small, it has to be repeated a large number of times. Not only are there several surfaces for each ray but, more important, a large number of rays must be traced from an object to the image plane in order that image quality may be judged. But that is not all, for having ray-traced an image for one set of parameters (refractive indices, lens shapes and separations, etc), the parameters (often 50 or more) must then be varied in an attempt to improve the image quality. Even when an acceptable design is found the search must

Optics and the Wave Nature of Light

continue in case an even better imaging system could have been found with a different initial set of parameters. This process is analogous to hunting for the deepest valley on the ocean bed. When you find a new record low it is still necessary to explore further in case there is an even deeper valley. As a consequence the design of most optical instruments is still based more on continuing improvements to an old specification than on any formal method of attack.

However, the recent advances in computer technology have caused a great change in this type of work. Their speed in performing arithmetic is clearly essential but there is another important aspect. Humans are very slow when dealing with numbers but have amazing ability when presented with pictorial information. Thus, if the computer is going to calculate the image of a test pattern of spots then it is very useful for the person concerned in the calculation to have the image displayed as a picture. This can in fact be done. The computer is able to make a television display of the coordinates of the image points. It is frequently useful for the designer, because of his experience, to be able to interrupt the calculation and re-direct the search for an optimum design. The computer programme can be instructed to characterise the image quality and continue searching routinely. But the designer frequently has a feel for the work and can short-cut the calculations. Man-machine interaction in this way utilises the power of computers (high computational speed) and the genius of the human brain. The effects of these new methods on the design of microscopes and other instruments is already becoming apparent.

Interference

We turn now from ray optics to the wave nature of light. Let us take the simplest case of a monochromatic plane-polarised wave, and ask the question, what happens when two such light waves cross each other? The answer is, in general, 'nothing', or at least 'nothing that we can detect'. We know this because in the real world light waves are continually crossing one another without any effect at all on our vision. However, under certain rather special circumstances effects due to the wave nature of light can be seen to give rather curious results. We call this phenomenon *interference*; we shall shortly outline what the special requirements are for interference to occur, but for the moment let us suppose that they are satisfied and discuss what happens when two light waves (which, for simplicity, we shall assume to have the same amplitude) intersec one another. We shall show that it is then possible for the two light beams either to reinforce each other or else to cancel each other out. At first sight this possibility of cancellation may seem rather extraordinary; but when you consider what type of process is going on when light is detected by the retina or anything else, it becomes understandable. The electric field of the light is exerting a force on electrons in the material comprising the detector. If the two beams light mentioned above are both incident on the detector then any one electron will experience two forces due to the electric fields of bot beams. Now assume that the two beams are plane polarized in the s plane. Depending on whether the two light beams are 'in step' or no these two forces on the electron can be equal and opposite, and hen cancel, or can reinforce each other, and hence add, or can add vectorially in any intermediate position. The two extreme possibilities are called *constructive* and *destructive* interference. Interference is quite general property of all wave motion. It is seen, for example, w sea waves reflect off breakwaters or sea walls, and interfere with th advancing waves to produce patches of still water (destructive) or \ turbulent 'broken water' (constructive). Another commonplace examp is the interference of sound waves — anyone who has listened to or chestras playing in certain older concert halls will know of 'dead spots', places in the auditorium where the music is inaudible due to destructive interference between the direct sound and that reflected off the walls or ceiling.

To go further we have to discuss the conditions which give rise to either constructive or destructive interference. This can be done either graphically or by using algebra, and we shall show each approach in turn. First, the graphical method.

The electric field strength (the displacement of the electromagnetic wave) varies in a simple harmonic manner with the frequency of the electromagnetic wave. (Simple harmonic motion was discusse in Chapter 1, pp. 27—34.) The resultant displacement or electric fiel strength when two such wave trains cross is found by adding the displacements of the two individual wave trains. This can be done graphically as is shown in Fig. 5.4. The case where the two wave trains have zero phase difference (a concept explained in Chapter 1) is shown in Fig. 5.4(a). They then interfere constructively. When the two wave trains are exactly out of step, that is when the phase difference is π, destructive interference occurs. This is shown in

Optics and the Wave Nature of Light

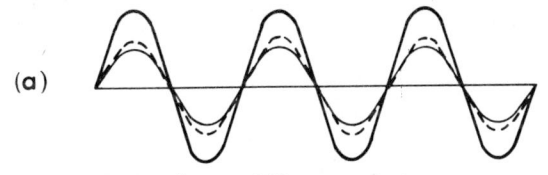

(a)

Zero phase difference between waves 1 and 2

(b)

Waves 1 and 2 are π radians out of phase

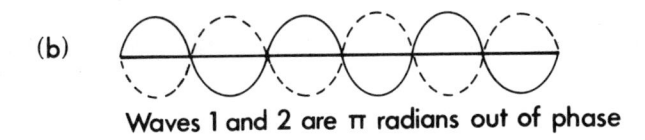

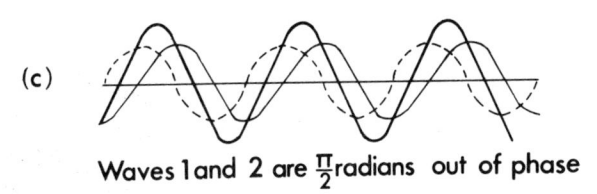

(c)

Waves 1 and 2 are $\frac{\pi}{2}$ radians out of phase

Fig. 5.4
Interference of two waves (light solid line and dashed line), showing the change in the resultant (heavy solid line) as the phase varies.

Fig. 5.4(b). A case of intermediate phase difference is shown in Fig. 5.4(c). Clearly, this graphical method would be tedious to use all the time, and it will simplify matters to use the following algebraic approach.

As we have seen in Chapter 1 a simple harmonic variation in displacement can be represented at a time t by

$$\text{displacement} = a \sin(\omega t + \phi)$$

where a is the amplitude, ω is the angular frequency, that is 2π times the frequency, and ϕ is a phase angle. The resultant displacement of two wave trains of the same amplitude but with a phase difference of δ is given by simply adding these two displacements together,

$$\text{resultant} = a \sin\omega t + a \sin(\omega t - \delta) \quad (5.2)$$

$$= \left(2a \cos\left(\frac{\delta}{2}\right)\right) \times \sin\left(\omega t - \frac{\delta}{2}\right) \quad (5.3)$$

(Equation 5.3 is derived from eqn 5.2 by using the following three trigonometric equalities

$$\sin(A \pm B) = \sin A \cos B \pm \cos A \sin B$$
$$\cos(A \pm B) = \cos A \cos B \mp \sin A \sin B$$
$$\cos^2 A + \sin^2 A = 1.$$

As an exercise, check this derivation and satisfy yourself that you understand it.)

Since $2a \cos\left(\frac{\delta}{2}\right)$ is a constant we see that eqn (5.3) can be interpreted as a simple harmonic wave of amplitude equal to $2a \cos\left(\frac{\delta}{2}\right)$, and of frequency ω again. Now we already know from eqn (1.43) that the phase difference δ is related to the path difference between the waves, x, say, by the equation

$$\delta = \frac{2\pi}{\lambda} x$$

Thus we see that if x is zero, the resultant amplitude is $2a \cos(0)$ = $2a$. This is constructive interference and the resultant is a wave whose amplitude is twice that of the original ones. On the other hand a path difference of $x = \lambda/2$ gives $\delta = \pi$ and a resultant amplitude of $2a \cos(\pi/2) = 0$. This is destructive interference. These results are consistent with what was got by drawing the waves (Fig. 5.4). Nothing new has been done except to introduce some algebra.

Such interference of two light waves can be observed using the apparatus shown schematically in Fig. 5.5. Light from a lamp is made monochromatic by a colour filter before passing through a narrow slit S. The wave nature of light causes it to spread out a little after passing through slit S. Slits S_1 and S_2 are equidistant from S and so waves reaching S_1 and S_2 are exactly in phase (no path difference between them). The light spreads out after passing through slits S_1 and S_2; the two wavefronts reaching a screen at a point P equidistant from S_1 and S_2 will interfere constructively

Optics and the Wave Nature of Light

because the waves are still exactly in phase. However, light from S_1 and S_2 reaching a point P', which is a distance y below P on the screen, will not be in phase because the distances $S_1 P'$ and $S_2 P'$ are not, in general, equal. If $S_1 P' - S_2 P'$ is $\pm \frac{\lambda}{2}$, $\pm \frac{3\lambda}{2}$, $\pm \frac{5\lambda}{2}$, etc., then the waves will interfere destructively. If $S_1 P' - S_2 P' = \pm \lambda, \pm 2\lambda, \pm 3\lambda$, etc., then the waves will interfere constructively. Thus the light intensity will vary from a maximum at P down to a minimum (when the path difference is $\frac{\lambda}{2}$ and then up to another maximum (when the path difference has increased to λ) and so on. Thus the screen is covered with a pattern of light and dark *fringes*. The distance y of the nth bright fringe from the centre P is given by calculating the path difference using geometry, and the result is

$$y = \frac{n\lambda D}{d} \qquad (5.4)$$

The distance D from slits to screen is always made very much larger than d, the separation of the slits, so that the spacing of the fringes can be easily observed.

Example: Red light of wave length 6600 Å from a narrow slit falls on a double slit with a separation between the centres of 0·02 cm. If the interference pattern is formed on a screen 1 metre away, what will be the separation between the fringes on the screen?

Using eqn (5.4) we see that the distance of the nth and $(n + 1)^{th}$ bright fringes from the centre of the pattern is given by

$$y_n = \frac{n\lambda D}{d}$$

$$y_{n+1} = \frac{(n + 1)\lambda D}{d}$$

respectively. Therefore the fringe separation is

$$y_{n+1} - y_n = \frac{\lambda D}{d}$$

$$= \frac{6600 \times 10^{-8} \times 100}{0 \cdot 02} \text{ cm}$$

$$= 0 \cdot 33 \text{ cm}$$

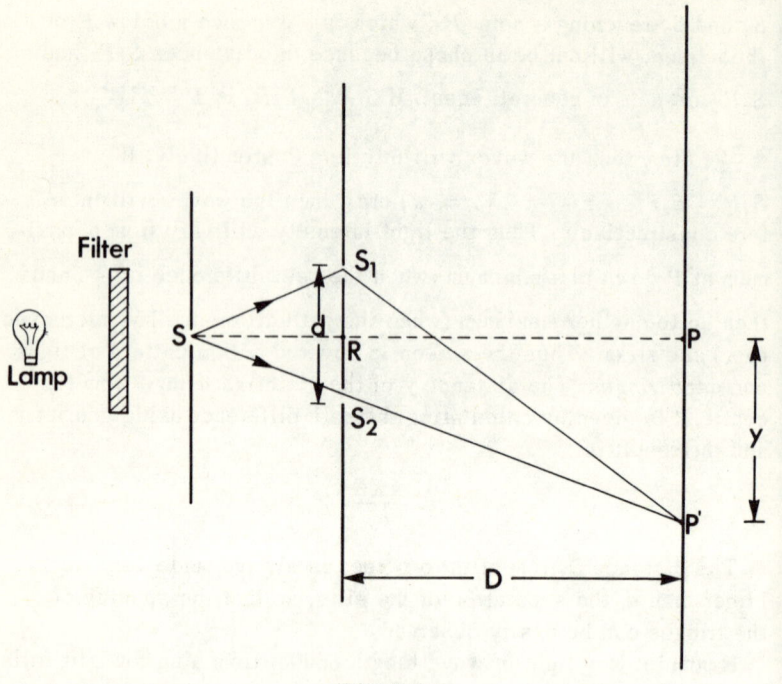

Fig. 5.5
Interference of light passing through two slits S_1 and S_2.

It is found that if the slit S is removed the fringe pattern is no longer visible to the eye, even though S_1 and S_2 are still illuminated. The reason for this is that the phases of the light rays getting through S_1 and S_2 fluctuate randomly at roughly 10^{-9} second intervals. This is because the light from the lamp originates from individual atoms, each of which only emits a wave train for about 10^{-9} second. Thus the light beam consists of a large number of uncorrelated wave trains each of which is about one foot long, as we pointed out in an earlier section. In the absence of slit S light from many different parts of the source will be passing through S_1 and S_2, and there is no correlation between the starting and stopping times of the wave trains in the different parts of the lamp. Consequently, although the waves *do* interfere on reaching the screen, their relative phases vary suddenly and at random so the fringes produced shift their positions in times of the order of nanoseconds. The eye is only capable of registering the time average because it has a much slower speed of response and thus sees a uniformly illuminated screen. The reason that fringes are seen if slit S is in position is not that such fluctuations do not occur but that the phase changes are the same for both S_1 and S_2 and hence the differences cancel out.

Light from slit S (and therefore from only a small region of the lamp) is called *coherent* light, that is light whose phase changes all occur at the same time. Lasers, which we will discuss later, are another source of coherent light. We stress again that interference is only seen to occur when the two interfering sources are coherent. This is why we do not usually see interference effects in the world around us. Most natural light sources are quite large and the wave trains they emit are not coherent with each other.

Interference is a huge subject with many applications, so the discussion here must necessarily be brief. However, there is one further class of interference phenomena which should be mentioned because it provides the only obvious example in nature where no special apparatus is needed to see the effects. In this class coherence is achieved even though light from a large extended source is used. An example is the colour seen in oil patches floating on water or on wet concrete or tarmac. Some light is reflected from the top surface of the oil and some from the water-oil boundary. These two components are brought together again in the eye of the observer. Destructive interference for particular wavelengths effectively removes these wavelengths from the light leaving the remaining light coloured. Coherence is achieved because any phase fluctuations in the incident light affect both reflected components and cancel out. The particular wavelengths removed, and hence the colours seen, depend on the thickness of the oil film and hence on the path difference of the two reflected wave trains.

A similar explanation accounts for the colours of soap bubbles. Also, the irridescent sheen of many insects such as beetles is due to a thin transparent surface coating in which the same effect occurs.

Example: Suppose a soap bubble has a purplish colour just before it bursts, and we want to know its thickness when this happens. We can calculate this if we assume that the purple colour is due to destructive interference of yellow-green light. The relevant formula here is that, for destructive interference at a wavelength λ, $\lambda = 2nd$ where d is the thickness of a film of refractive index n. Inserting $\lambda = 5 \cdot 3 \times 10^{-7}$ m, say, for yellow light, and $n = 1 \cdot 33$ for water, we find $d = \lambda/2n = 2 \times 10^{-7}$ m or 2000 Å.

Diffraction

Light does not always travel in straight lines but bends round obstacles in its path. This behaviour is common to all waves. The magnitude of the effect depends on the wave length of the waves and the obstacle size. The effect becomes large when the obstacle size is comparable with the wave length. This is why the slit widths in Fig. 5.5 must be very narrow; the bending of the light waves at S_1 and S_2 is the only reason they are able to meet and recombine on the screen.

This behaviour of waves at an obstacle (or an aperture) is called *diffraction* and its effect is that differently shaped obstacles and apertures produce various *diffraction patterns* after the light has bent round or through them. These diffraction patterns are the result of interference between portions of the wave which have travelled along varying paths around the obstacle or through the aperture.

Two typical diffraction patterns are shown in Fig. 5.6. Monochromatic light from a distant point source passes through a hole in a mask and falls on a screen. The general appearance of the light patch on the screen for a large hole (several cm in size, say) is a uniformly bright central region with fringes of varying intensity at the geometrical limits of the patch. The important point is that some light reaches the screen outside the geometrical limits. The scale of the fringe separation and excursion beyond the geometrical limits is such that it is hard to see with the naked eye. This is the reason why rectilinear propogation is such a good approximation in everyday circumstances.

As the hole size is made smaller and smaller the light pattern on the screen changes (Fig. 5.6b). Eventually a central patch of light with light and dark fringes at the edges is seen. Calculation shows that roughly 84% of the light intensity falls in the central peak.

This variation of intensity in the light patches can be understood if the phase difference between radiation arriving from different parts of the aperture is taken into account. The dependence of the size of the diffraction pattern (e.g. distance from centre of pattern to the first dark fringe in Fig. 5.6b) on the wave length and hole size will be simply derived. The result is very useful in all diffraction problems. The only change which is introduced into the formula when the hole shape is altered is to multiply the result by a numerical constant. So for any wave motion and all obstacles

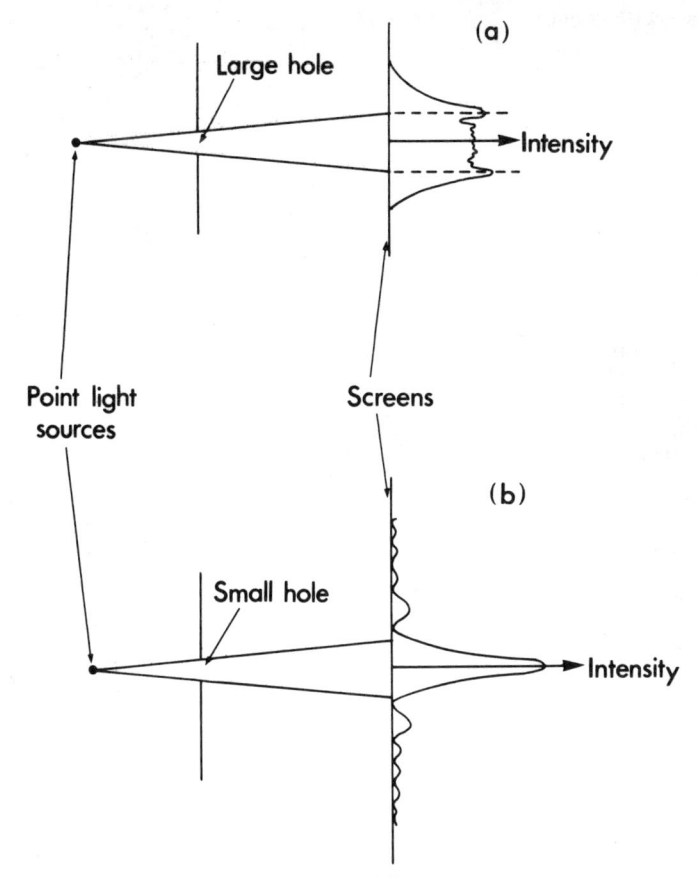

Fig. 5.6
Light passing through an aperture bends slightly resulting in interference between wave fronts arriving at a screen. (a) and (b) show the resultant fringe patterns due to holes which are respectively large and small compared with the wavelength of light.

and apertures we can get an understanding of what will happen by using our simple formula.

In order to derive this useful result consider that the pattern of Fig. 5.6(b) is produced by a plane wave of monochromatic light of wavelength λ falling on an aperture of width W and eventually onto a screen a distance D further on (Fig. 5.7). The wavefront is parallel to the slit. Thus the phase of the wave is constant across the slit (e.g. a crest everywhere along ABC). Imagine Fig. 5.7 to be drawn with θ chosen such that $BF = \frac{\lambda}{2}$ and $CE = \lambda$.

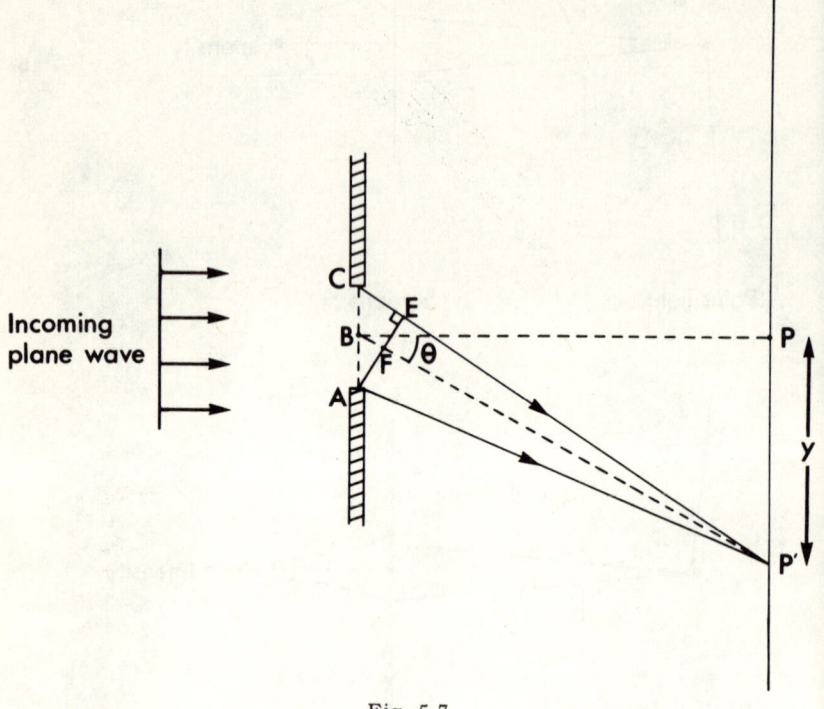

Fig. 5.7
Diffraction of light at an aperture.

At point P, the geometrical centre of the pattern, the light intensity will be a maximum. At points on the screen below P the light intensity will fall off due to an increasing amount of interference between waves arriving out of phase. The ultimate is reached at point P' where the wave arriving from point B is exactly out of phase $\left(\frac{\lambda}{2}\text{ path difference}\right)$ with the wave arriving from point A. The two contributions therefore exactly cancel. Exactly similar comments hold for waves starting from points immediately above A and B. By repetition of the argument one sees that there is complete cancellation of all the waves arriving at point P'. P' therefore corresponds with the first dark fringe outside the main intensity peak (Fig. 5.6b). The condition that θ must satisfy for this dark fringe is

Optics and the Wave Nature of Light

$$CE = W \sin \theta = \lambda$$
$$\therefore W\theta = \lambda \qquad (5.5a)$$

where use has been made of the approximation $\sin \theta = \theta$ for small values of θ. In the same approximation

$$\theta = \frac{y}{D}$$

$$\therefore y = \frac{\lambda D}{W} \qquad (5.5b)$$

Equation (5.5) contains the most important information about diffraction, namely its dependence on wave length and aperture size.

Example: Such a diffraction pattern is quite easy to see if we prick a sheet of black card with a needle and then squint through the tiny hole at a monochromatic light source (a sodium vapour street lamp, say) some distance away. For a typical sewing needle the aperture size will be about 0·2 mm; inserting this value for W, and putting $\lambda = 6 \times 10^{-4}$ mm in eqn (5.5a), we find $\theta = \lambda/W = 6 \times 10^{-4}/2 \times 10^{-1} = 3 \times 10^{-3}$ radians which is approximately one sixth of a degree. This means that if we squint at the pinhole from a distance of 100 cm the dark fringe will appear as a ring of apparent radius 3 mm. (Draw a sketch to convince yourself that this is so.)

Another topical example of diffraction concerns an experiment left on the Moon by the Apollo astronauts. This consisted of a mirror to reflect back laser light that was shone from the earth. The light was sent out by using a telescope at the McDonnel Observatory, Texas, 'backwards', that is by putting the light source at the eyepiece and letting the light emerge from the objective. The telescope diameter was 107". If we assume that the transmitted light beam spreads out because of diffraction we can calculate the size it will have when it strikes the Moon. We use eqn (5.5b) with $\lambda = 5 \times 10^{-5}$ cm, D = distance to the Moon = 4×10^{5} km, W = 107" = 272 cm. Then the size of the light patch on the Moon = $y = \lambda D/W = 5 \times 10^{-5} \times 4 \times 10^{5}/272 = 74$ m. In fact the observed size of the light patch was rather larger than this owing to turbulence in the earth's atmosphere.

There is a very pleasing and economical theorem, named after its discoverer Babinet, which states that in a diffracting system, if all the holes are replaced by opaque material and all the opaque material replaced by transparent material (i.e. a hole) then the diffraction pattern will be unaltered. We have been considering diffraction by a hole in a screen. Exactly the same diffraction occurs round a disc of the same size and shape as the hole. Hold out a penny towards a distant street lamp, preferably a monochromatic source such as a sodium lamp. It should block out the lamp if no diffraction occurs. However, provided it is not brightly lit where you stand you will observe a spot of light when looking towards the *centre* of the penny. Diffraction is occurring; light waves are bending around the rim of the coin and constructive interference occurs as they recombine on one's retina.

Another simple result about diffraction that is proved in more advanced textbooks is that a randomly placed array of circular holes (or discs) yields the same sort of diffraction pattern as a single hole (or disc), only it is much more intense. If the holes are not all exactly the same size but vary slightly, the result still applies but the diffraction pattern is governed by the average of the hole sizes, and also it tends to get very fuzzy at the edges.

An interesting application of this is a method of measuring the size of the red blood corpuscles of a person. These corpuscles are remarkably uniform in size for any one individual, so a fairly clear diffraction pattern is obtained. If a drop of blood is smeared on a glass plate and allowed to dry, the opaque corpuscles will behave as diffracting discs to incident monochromatic light. Measurements on the resultant diffraction pattern allow the mean corpuscle diameter to be estimated. This device is sometimes called an *eriometer*.

A transparent object with a number of diffracting elements arranged in an ordered array, rather than randomly, is called a *diffraction grating*. Such gratings usually consist of a number of parallel equidistant slits or grooves carefully ruled on a glass or metal plate. The slits are very close together and their spacing is accurately known. The main use of these gratings is in the measurement of wavelength in light spectra, and the method of doing this is illustrated schematically in Fig. 5.8. Monochromatic light passes through a slit S, parallel to the grating slits, and a device called a collimator focusses this light into a parallel beam. For simplicity we consider the case where the grating is perpendicular to the incident light. Then the plane wave fronts are parallel to the grating and light reaches all grating slits with the same phase. Light is diffracted out in all directions by each slit. All the waves from the different slits are in phase when they continue to progress in the original direction AB so in this direction constructive interference occurs. When all the waves progressing in this direction are brought

Optics and the Wave Nature of Light

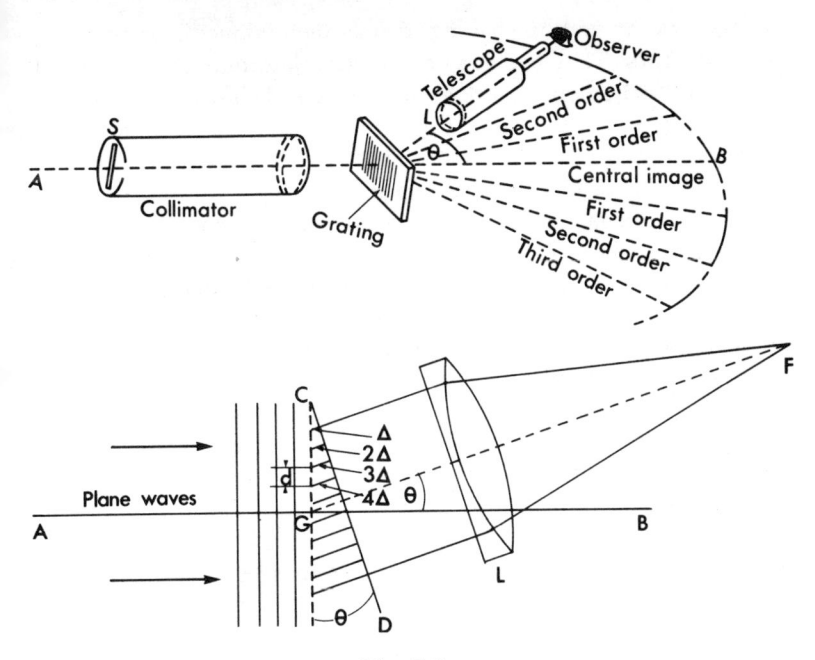

Fig. 5.8
Schematic diagram of a diffraction grating as used in a spectrometer for measuring the wavelength of light.

together (by a lens, or simply by looking along this line) they interfere constructively to produce a bright central image of the slit S. But it is also possible to get constructive interference in a direction such as GF which makes an angle θ with AB. The necessary condition is that the path difference Δ between light beams coming from successive slits is one, two, three etc. wavelengths. Along the line CD perpendicular to GF, light from the first slit below C will be a distance Δ ahead of light from C; that from the second slit, 2Δ ahead, etc. The geometry of Fig. 5.8 shows that

$$\sin \theta = \frac{\Delta}{d}$$

Using the condition that for constructive interference

$$\Delta = n \lambda \text{ where } n \text{ is an integer}$$

gives the result

$$n \lambda = d \sin \theta \tag{5.6}$$

where d is the grating spacing, that is the distance between successive slits. Thus besides constructive interference in the direction AB there will be constructive interference when $n = 1$ at $\theta = \theta_1$, $n = 2$ at $\theta = \theta_2$ etc. These images of the slit are referred to as the first order image, second order image, etc. They appear, of course, on both sides of the straight-through direction.

If the incident light is not monochromatic but contains several discrete wave lengths (e.g. light from a fluorescent lamp) then the central image consists of all the colours superimposed. As θ is increased from zero, so first order images of the slit will be encountered, the colours ranging from the blue end of the spectrum through to red as θ increases. Continuing to increase θ brings the second order images into view, and so on. Thus by measuring the angles θ at which bright images of the silt can be seen, one can calculate the wavelength of the light by substitution in eqn (5.6).

Example: Suppose light of wavelength 667 nm is incident normally upon a diffraction grating ruled with 6000 lines per cm. At what angles will the first and second order images be seen?

For the first order, $n = 1$ and so $\sin \theta_1 = \lambda/d = 667 \times 10^{-9} \times 6 \times 10^5 = 0.4000$. Hence $\theta_1 = 23.6°$.

For the second order, $n = 2$ and so $\sin \theta_2 = 2\lambda/d = 2 \times 667 \times 10^{-9} \times 6 \times 10^5 = 0.8000$. Hence $\theta_2 = 53.1°$.

Resolving power: the limit set by diffraction

In this section we are going to demonstrate how the resolving power of optical instruments is limited, not by the geometrical optics of their construction but by the wave nature of light itself. The reason this is so is that the apertures of optical instruments, although usually large compared with the wavelength of light, still allow some diffraction to occur. That is, the light waves bend slightly on passing through the apertures and instead of producing sharp geometrical images cause their edges to become blurred and fuzzy. This clearly limits the possible power of the instrument. The two examples we shall discuss are the optical microscope, and the compound eyes of insects.

The purpose of the microscope is to see fine details which are not visible to the naked eye. Its primary function is to yield an image in which small details are optically resolved. As will be shown, magnification as such is easily achieved and is not a limiting factor on microscope performance. *Resolving power* is defined as

Optics and the Wave Nature of Light 131

the smallest distance between two point-like objects which can just be seen as separate with the microscope. Resolution is achieved if the visual angle between the objects themselves, or their magnified images, is large enough that light rays from either point will fall on different receptors in the retina. For the unaided human eye the minimum angle which can be resolved is about one minute of arc or 0·29 milliradians. Thus two point objects viewed from 25 cm (the least distance of distinct vision) can just be resolved by the naked eye if their separation is 0·07 mm or 70 μm. This can be improved on, by using a compound microscope, to about 0·2 μm. We shall show that this cannot be bettered by using visible light because of diffraction effects. (However substantially better resolution is achieved with an electron microscope because of the shorter wave length involved, see Chapter 6.)

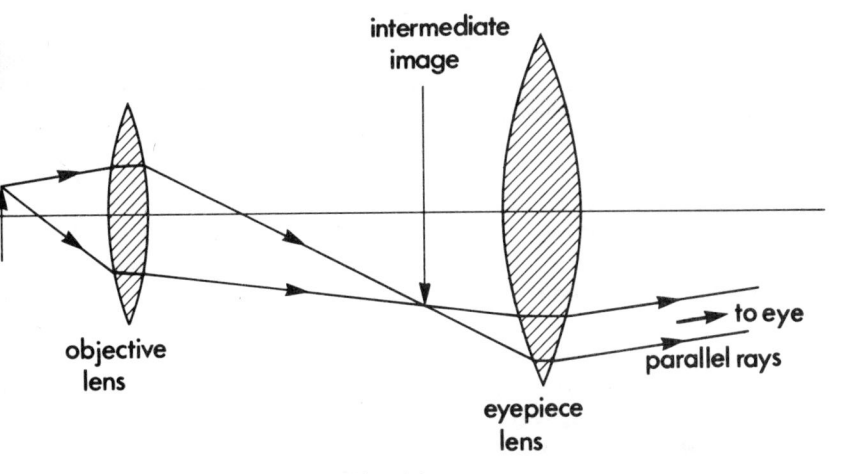

Fig. 5.9
Image formation in the compound microscope.

The path of one bundle of rays in the microscope is shown in Fig. 5.9. Both the objective and the eye-piece are represented schematically by a single lens. When the microscope is focussed on an object O, the eye of the observer is relaxed as if viewing a distant object. The objective lens forms an inverted, enlarged, real image of the object in the lower focal plane of the eye-piece. This intermediate image is viewed with the eye-piece acting as a magnifying glass. The diverging bundles of rays are changed to parallel bundles by the eye-piece. Refraction in the eye then leads to an image on the concave surface of the retina. From this description of the role of the eye-piece, it can be seen that it is the objective which limits the resolving power of the microscope. Any details which are not separated in the intermediate image cannot be made visible through magnification in the eye-piece. This can be seen from another point of view. A point on the object emits a spherical wave front of light. Only a limited cone of this light can be collected by the objective. Thus only limited information about the object is collected. The eye-piece cannot yield more information than originally collected by the objective.

A consequence of the light from the object passing through an aperture (in this case, the objective lens) is that diffraction occurs. A point image of a point object is not formed, and it must be stressed again that this is so even for a perfect, aberration-free lens. The image which is in fact formed is shown in Fig. 5.10(a). It consists of a central bright disc surrounded by alternately dark and bright rings. Once again about 84% of the total light intensity falls within the central bright disc. The radius of the disc, that is the distance from its centre to the first dark fringe, is

$$r = \frac{0 \cdot 61 \lambda}{n \sin u} \tag{5.7}$$

where n is the refractive index of the medium between the object and the objective lens, and u is the half angle of the cone of light accepted by the objective. The product $n \sin u$ is called the *numerical aperture*.

A similar diffraction effect occurs with astronomical telescopes which form, not point images of stars but the same diffraction pattern we have been discussing. The numerical factor of 0·61 in eqn (5.7) was indeed first calculated in 1834 by Sir George Airy, later Astronomer Royal, and the bright central patch itself is known as the *Airy disc*.

The limit to resolution arises from the size of the Airy disc. The image of two separate points is two Airy disc patterns. Imagine the two object points to be moving closer together. So of course will their images. However, the Airy discs, because of their size, will overlap before the two objects coincide. In fact when the centre of one Airy disc coincides with the first dark ring of the other Airy disc pattern (as in Fig. 5.10b) it is accepted that the two images can no longer be seen as separate by the eye.* The distance apart

---
* The exact separation at which the two Airy discs cannot be seen as separate varies from individual to individual. It is definitely possible to decide that two images are present at separations of the centres smaller than the radius of the first dark ring but the above is the conventional definition.

Optics and the Wave Nature of Light

(a) Airy Disc of a single point

(b) Airy Discs of two closely adjacent points

$d = r = \dfrac{0.61\,\lambda}{n \sin u}$

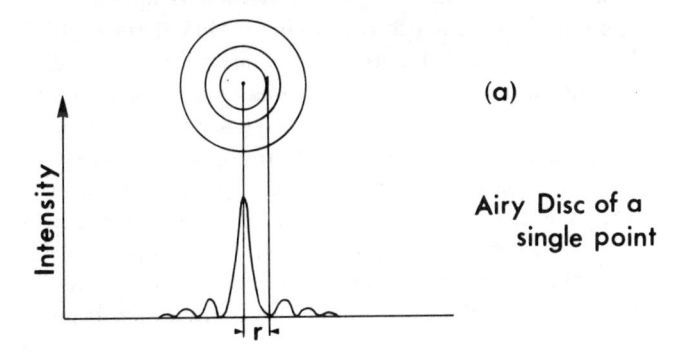

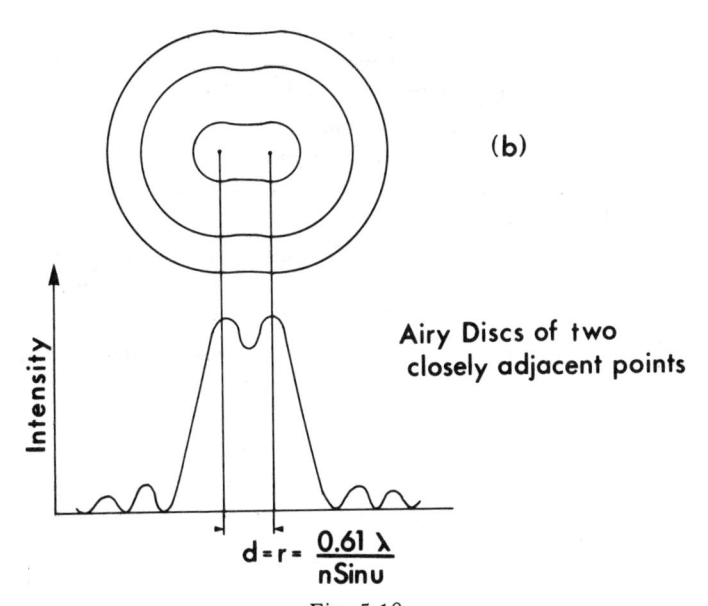

Fig. 5.10
Airy disc diffraction patterns formed by (a) a single point object and (b) two adjacent point objects.

of the two object points is then the absolute limit of resolution. Thus the resolving power d is given by

$$d = \frac{0\cdot 61\,\lambda}{n\,\sin u} \qquad (5.8)$$

The resolving power can be improved by increasing $n \sin u$. Now increasing $\sin u$ corresponds to increasing u also. This angle, u, depends on the ratio of objective lens diameter to object distance. On trying to increase u great care has to be taken to avoid lens aberrations since most of the light rays are well off-axis. The objective lens is usually a complicated arrangement of several lenses, the whole combination having been carefully designed according to the requirements outlined in an earlier section. An increase in n can be achieved by filling the region between the object and the objective lens with oil — a technique known as *oil immersion*. Decreasing the wave length of the light also improves resolution.

A knowledge of the resolving power of a microscope is useful in selecting the appropriate eye-piece. As has been pointed out, even the best eye-piece cannot reveal details which are not resolved in the intermediate image formed by the objective. The eye-piece's purpose is to act as a magnifier, making optically separate details of the intermediate image recognisable to the eye. In order to do so it must present these details at an angle greater than one minute of arc because of the limit of angular resolution of the human eye. Nothing is to be gained by using a higher magnification eye-piece than that required to achieve this. Weaker eye-pieces do not permit recognition of all the details in the intermediate image while higher magnification eye-pieces will not disclose any new image details. In certain cases (particle counting, absorption measurements) over-magnification can relieve eye-strain during long periods of work.

As an example of what optical microscopes are capable of, suppose that the angle u in eqn (5.8) is $90°$, so that $\sin u$ has its maximum value of one. If n is 1·8, as is possible with some oils, and λ is 300 nm (the near ultraviolet) the resolving power is given by eqn (5.8) as $\frac{0·61 \times 300 \times 10^{-9}}{1·8 \times 1} = 10^{-7}$ m $= 0·1 \mu$m. This represents the smallest distinguishable separation between objects, without the use of utterly novel techniques (see Chapter 6). It was in fact the best resolving power available until about 25 years ago.

We turn now to our second example of diffraction effects, in the compound eyes of insects. These highly specialised eyes are the main organs of vision in adult insects, consisting of aggregations of separate visual elements called ommatidia. There is great variation in the number of ommatidia making up a single compound eye. In the worker class of one type of ant there is only one, in that of

Optics and the Wave Nature of Light

another type there are six to nine; and in large dragonflies, up to 28 000. The compound eye is a less perfect optical instrument than the vertebrate eye because there is no general focussing device present. Consequently, each ommatidium forms its own image of a small portion of the field of view and all combine to give a mosaic effect of the whole.

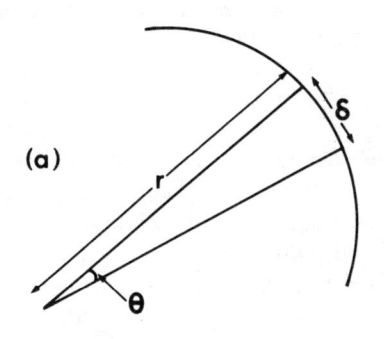

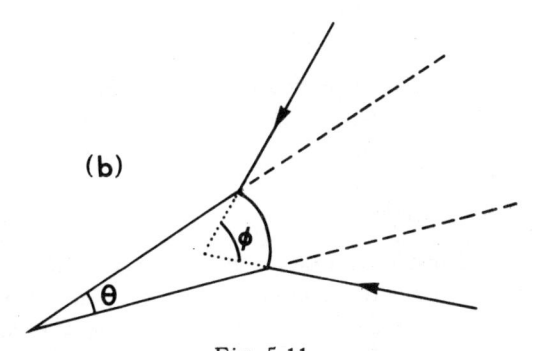

Fig. 5.11
(a) A single ommatidium cone. The eye surface is assumed to be a sphere of radius r. The entrance to the ommatidium is of size δ. (b) The dashed lines show the geometric light cone of angle θ. Due to diffraction, light rays within the larger cone of angle ϕ can enter the ommatidium.

There are two main factors which affect the sharpness of the overall image (the visual acuity). The first is the opening angle θ of the ommatidia (see Fig. 5.11a) which can be envisaged as cones emanating from the centre of a sphere. If the sphere (eye) radius is r and the typical dimension of the entrance is δ then the opening angle θ

is approximately δ/r. One might imagine that the visual acuity of the eye could be increased by having more ommatidia, of smaller θ. There is however a limit, set by diffraction, to how small θ can usefully be made. Light shining through the front of the eye is diffracted so that it spreads out beyond the limit allowed by geometric optics. Thus light from a cone of opening angle ϕ, rather larger than θ, can enter the ommatidium because of this diffractive bending. Rewriting eqn (5.5a) gives

$$\phi = \frac{\lambda}{\delta}$$

Thus as δ, the size of the ommatidium entrance, gets smaller, and therefore as θ gets smaller, ϕ will become larger. If θ gets too small ϕ will have increased so much that the fields of view of neighbouring ommatidia overlap and there is no further gain in visual acuity.

Thus for a given radius of eye there is an optimum value of the ommatidium entrance size which results in the best visual acuity. A reasonable way to estimate this optimum size might be to choose it so that θ and ϕ were roughly equal, i.e. $\theta = \phi$. This would imply that $\frac{\delta}{r} = \frac{\lambda}{\delta}$ and so

$$\delta = \sqrt{\lambda r} \qquad (5.8)$$

Measurements on actual insects show that the dimensions δ and r are in fact roughly related by eqn (5.8) so that evolutionary processes do seem to have resulted in optimised eye dimensions.

The polarisation of light

It was mentioned in the section on the nature of light that the electric vector is always perpendicular to the direction of motion. For example, if we choose axes as in Fig. 5.1 the electric vector must lie somewhere in the X-Y plane, for instance along the Y axis as drawn. In the topics discussed so far this orientation of the electric field vector has not been of importance. The only exception arose in the discussion of interference, when we insisted that the two interfering light waves each be *plane polarised* in the same plane, that is have their electric vectors pointing in the same direction. This was necessary in order for interference to occur, and in fact for coherent sources it is always the case. Otherwise we have tacitly assumed that the light waves we are dealing with are *unpolarised*, that is consisting of many wave trains all with their electric vectors

Optics and the Wave Nature of Light

pointing in quite arbitrary directions in the X-Y plane. We have further assumed that there is no relation at all between the polarisation direction (the electric vector direction) of one wave train, and that of all other wave trains. This assumption is generally true if we are dealing with light from natural sources such as hot filaments or gases — each wave train comes from a different atom or molecule and there is no reason to suppose that any particular polarisation direction is preferred. In this section we discuss phenomena for which the direction of vibration of the electric field *is* of great significance.

We first discuss the various types of polarisation that a light wave can possess. Then we consider how this polarisation can be produced and detected, finally mentioning some special properties and applications of polarised light.

To proceed, we remind the reader that any vector in the X-Y plane can be resolved into two components, a vector parallel to the X axis and another vector parallel to the Y axis.

In Fig. 5.1 the orientation of X-Y axes is quite arbitrary. Thus a plane polarised electromagnetic wave travelling along the Z axis could have its electric field vector (E vector for short) oscillating at an arbitrary direction in the X-Y plane. This vector can then be considered as the resultant of its component vectors along the X and Y axes. Clearly when the resultant E vector is zero so also are its components along the X and Y axes. Similarly when the E vector is a maximum so will its components be a maximum. In other words the components along the X and Y axes are in phase with each other, and with the E vector itself. Thus a plane polarised electromagnetic wave is equivalent to two plane polarised electromagnetic waves whose electric field vectors are in phase and perpendicular to each other.

It is interesting, and will shortly prove relevant, to consider the wave motion produced by superimposing two waves which are *not* in phase but whose electric vectors are perpendicular. The resultant E vector then varies in time. It can be shown that, in general, the tip of the E vector moves round an ellipse in the X-Y plane (a circle being one special case of an ellipse). If the reader finds this resultant elliptical motion hard to visualise perhaps the following analogy will help. Consider a ball suspended by a string, free to oscillate in either the X direction or the Y direction. The most general motion, a combination of both X and Y oscillations out of phase is an ellipse — try it.

Thus as light moves down the Z axis the electric vector can move either to and fro along a straight line, around an ellipse or a circle. All of these cases are physically possible and there is a terminology to describe them. Light is *linearly* or *plane* polarised when the electric vector oscillates on a straight line (Fig. 5.1). When the end of the electric field vector moves round an ellipse, the light is *elliptically* polarised. When the end of the electric field vector moves round a circle, the light is *circularly* polarised. If the end of the electric vector, when we look along the light as it travels away, goes round in a clock-wise direction, it is called *right-hand* circular polarisation. *Left-hand* circular polarisation corresponds to the opposite direction of rotation (though we should warn the reader that these conventions regarding handedness are not universal). As we have already said, so-called 'unpolarised light' has a polarisation which is rapidly changing. Each atom in the source emits an electromagnetic wave train for $\sim 10^{-9}$ sec and then stops. Thus a light beam consists of a number of wave trains each about a foot long and having a particular polarisation. The polarisations of the different wave trains are not correlated in any way. Thus the polarisation of the total light beam keeps changing every 10^{-9} sec or so. This is too rapid to detect and so the light is called unpolarised.

Since light from most sources is unpolarised the question arises of how polarised light is produced. Given plane polarised light, elliptically polarised light can be produced by combining two beams of perpendicularly plane polarised light, of different intensity and relative phase. Circular polarisation is a special case of elliptical polarisation. Thus the key problem is how to produce plane polarised light, since from such light we can make the other types of polarisation.

Fortunately there exist materials whose light absorption is different for light polarised in different directions. A beam of unpolarised light incident on such a material would have those of its wave trains with one particular direction of polarisation strongly absorbed while those polarised at right angles to this direction would pass through the material without much attenuation. Wave trains polarised at an intermediate angle would have the component along the 'easy' propagation direction passed without much attenuation while the remaining component would be strongly absorbed. Calcite (or Iceland spar) and tourmaline are two naturally occurring minerals which have this property of passing one particular plane of polarisation. Better known perhaps, and much cheaper, is the commercially produced material, polaroid.

Optics and the Wave Nature of Light

The better quality sunglasses are made from polaroid, and its action in polarising light can be demonstrated by taking two pairs of such sunglasses and looking through the two together. When the spectacles are directly superimposed the field of view is simply a little darker than usual (since quite a lot of the light has the wrong polarisation direction and is stopped). If one spectacle is now rotated with respect to the other, the field of view will darken until, after turning through 90° it becomes essentially black. This is because each lens lets through only plane polarised light so when their axes of polarisation are at right angles no light at all can get through (apart from a small amount due to imperfections in the lenses).

Besides its production by transmission through certain special materials, plane polarised light can also be produced by reflection from, for example, a glass or liquid surface, and by scattering light from small particles. Consider light incident on the reflecting surface at such an angle i that the reflected and refracted beams are at right angles (Fig. 5.12). This happens when

$$i + r = 90°$$

where r is the angle of refraction. If the light is incident from air onto a material of refractive index n then Snell's law states that

$$\sin i = n \sin r$$

Eliminating r from these two equations gives

$$\tan i = n \tag{5.9}$$

The value of i given by eqn (5.9) is called Brewster's angle. Light incident at Brewster's angle will all be refracted, *if it is polarised in the plane of incidence* (the plane of the drawing in Fig. 5.12). It is not possible for any of it to be reflected. This is because it is the motion of electric charges in the reflecting material, induced by the incident electromagnetic wave, which generate the reflected beam. It can be seen that an oscillation in the plane of the paper cannot give rise to a reflected beam because its electric vector (polarisation direction) would be parallel to the direction of motion, and since we know that light is a transverse wave, this cannot be. Thus, the reflected beam can be polarised perpendicular to the plane of the paper but not in the plane of the paper.

This property is ingeniously made use of by the action of polaroid sunglasses in eliminating 'glare'. Most dazzle effects in bright

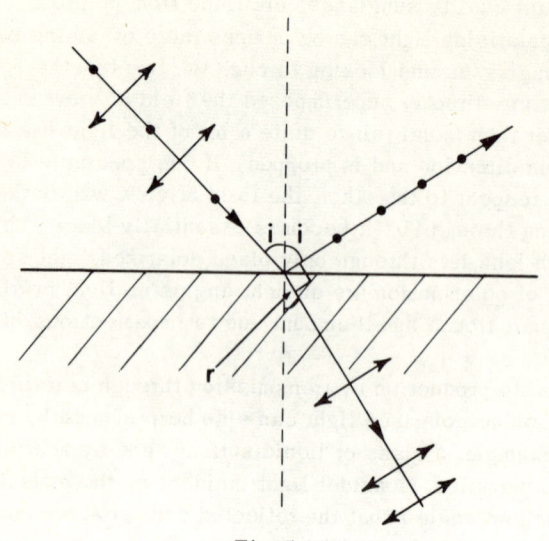

Fig. 5.12
Polarisation of light is achieved by reflection at Brewster's angle.

sunlight are due to the light being reflected off sheets of water, say, into one's eyes. The plane of incidence is then a vertical plane and the reflected light will be predominately plane polarised perpendicular to this, that is horizontally. So if the sunglasses have their axis of polarisation vertical a lot of this reflected light will be stopped, though not all of it because not all the reflections occur at, or even very close to, Brewster's angle. For example, Brewster's angle for water (n = 1·33) is about 53°, and since at latitude 45° the sun is at this altitude, or higher, for a great deal of a midsummer day such sunglasses are clearly very useful. It is fairly easy to show that they are not so useful at latitudes above the Arctic Circle. Try to draw a sketch to convince yourself of this.

Polarisation can also be produced by the scattering of light. It is well known that daylight arriving at the earth's surface is partially plane polarised. To demonstrate this fact, look at a patch of blue sky through polaroid sunglasses, and slowly rotate the spectacles in front of you. You will see the sky alternately darkening and brightening as the polarised light is stopped and transmitted respectively; notice that the degree of darkening depends on the direction in which you are looking. This exercise is not entirely pointless since there is now much evidence that some animals use this polarisation effect as a way of navigating. In particular, bees which unlike humans are directly sensitive to, and can distinguish between, various directions of polarisation are believed somehow to move from flower to flower, and return to the hive, by this method.

This polarisation of light from the blue sky is due to the sunlight being scattered off extremely small particles, such as dust, ice and salt crystals, in the upper atmosphere. (Such scattering not only polarises the light, as we shall see; it also colours it since the shortest wavelengths are preferentially scattered. This indeed is why the sky appears blue, and also why a setting sun is red — the blue part of its spectrum has been lost due to scattering in the thick layers of air.)

It can be seen how polarisation on scattering can occur when Fig. 5.1 (p.111) is consulted. Suppose we resolve the light beam into two components, polarised in the X and Y directions respectively. Let us consider whether both these components are equally capable of being scattered in the Y direction. The X component can be, since its plane of polarisation will still be perpendicular to its direction of travel. The Y component, however, cannot be, since such scattering would result in the oscillations of the electric vector, that is the plane of polarisation, being along the direction of travel. We have already seen that this is not allowed, light being a transverse wave motion, so scattering of the original unpolarised beam in the Y direction results in a completely polarised beam. Other directions correspond to partial degrees of polarisation.

Just as there are substances whose coefficient of absorption is different for different planes of polarisation so there are materials whose refractive index is different for different planes of polarisation. As an example, consider a crystal of long molecules which are in extended chain form, i.e. stretched out straight and parallel to each other. Suppose that the electrons of the molecule respond more easily to the oscillating electric field of light in the direction parallel to the axis of the molecules than at right angles to this direction. Then the response to plane polarised light will depend upon the orientation of the plane of polarisation. Call the direction of the axes of the molecules the *optic axis*. Then the refractive index of the crystal for light plane polarised perpendicular to the optic axis is different from the refractive index for light plane polarised parallel to the optic axis. That is, the two components are transmitted with differing velocities. It is interesting to consider what happens to light which is plane polarised at an intermediate angle, say 45° to the optic axis. This light can be resolved into two beams polarised at right angles to each other, and we may as well take the optic axis as one of these directions. The two beams are initially in phase but because of their different velocities they get progressively more out of phase as they travel through the material.

On leaving the crystal the light can be passed through a polariser, again set at 45° to the optic axis. The two components will recombine and, being now polarised in the same plane, can interfere. The two are certainly coherent (they were in fact a single wave train) so for some particular wavelength the phase difference will be such that destructive interference occurs. The result is that these wavelengths are effectively removed from the light which then appears coloured.

This property of appearing coloured in thin section, when viewed in polarised light, is common to a wide variety of substances, particularly minerals. It is due to the two different refractive indices, along and perpendicular to the optic axis, and is known as *birefringence*. The two refractive indices (often called the *ordinary* and the *extraordinary* refractive index to distinguish these materials from common isotropic substances) usually depend slightly on the wavelength of the light. This is the reason that perfect destructive interference occurs only for one or two specific wavelengths.

The technique of studying birefringence in the so-called 'polarising' or 'petrological' microscope is a very powerful one of some complexity. The reader will probably find it worthwhile to look at a few thin sections of common rock-forming minerals in such an instrument — the technicolour detail revealed is staggering.

Many materials are known which rotate the plane of polarisation of light passing through them. They are known as *optically active* substances. Many optically active solids are composed of molecules that are asymmetric, that is which can exist in both a left-handed and a right-handed form. Most sugars are optically active; in fact their activity is so striking that the names dextrose and laevose refer to the right- and left-handedness of their respective molecules. Such materials are capable of rotating the plane of polarisation of incident light by quite significant amounts, in a direction which is right- or left-handed depending on the particular substance involved. This rotation can easily be demonstrated by placing a thin sheet of material (quartz is quite a good sample for this demonstration) between crossed polaroids, that is polaroid sheets with their axes of

Optics and the Wave Nature of Light

polarisation set perpendicular to one another. The field of view is seen to brighten as the plane of polarisation is rotated, the actual angle of rotation being determined both by the substance itself and by its thickness.

The optical activity, when due to asymmetric molecules, is still observed in solution. Thus sugar solutions are optically active, a fact which is the basis of several commercial instruments (*saccharimeters*) for determining the concentration of syrups.

Example: The specific rotation (angle by which the plane of polarisation is rotated by a one decimetre column of solution of concentration 1 gram per cc) of dextrose is $+52 \cdot 7°$ and that of fructose is $-133°$. The + and − signs indicate that the direction of rotation is right- and left-handed respectively, using the convention that we mentioned above in discussing circular polarisation. Thus the degree of rotation caused by a mixture of twenty grams of dextrose and ten grams of fructose dissolved in a litre of water and poured into a four cm diameter glass tube is found by calculating the concentrations, finding the corresponding rotations caused by each solution separately, and adding the results. The answer is $-2 \cdot 19°/+2 \cdot 19°/-21 \cdot 9°$. (One of these answers is correct — check the working to find which.)

Laser light

One of the most significant developments in optics in the last few years has been the laser. It is important not only in the field of optics but in practically every area of physics since it can be utilised to make measurements of great interest. A laser is a source of high intensity, monochromatic, coherent light. In this section it will be explained how light from lasers differs from ordinary light and why this difference occurs. Then some of the many uses of lasers will be briefly described.

Laser light differs from ordinary light because it is more intense, directional, monochromatic and coherent. It has already been mentioned that the light emitted by ordinary sources consists of a series of wave trains, each about one foot long. These wave trains follow one another in time in a random, uncorrelated way. In contrast to this the laser effectively emits a continuous train of waves. Thus the waves of laser light are coordinated in both space and time in a way which is analogous to a column of soldiers marching perfectly in step. Again light from ordinary sources is far from monochromatic even if a single spectral line is selected by means of filters. Typically, the light from even the best low-pressure gas lamps is spread out over a band of frequencies of about 1000 MHz (10^9 cycles per second). Light from a gas laser can be confined to a band of 1 MHz. This spread in frequency corresponds to a certain spread in wavelength; an ordinary vapour lamp, with a wavelength near 600 nm, say, has a spread in wavelength of one part in 10^6; for laser light the spread is one thousandth of this.

The reason for this chromatic purity and coherence can be seen if the mechanism of the laser is very briefly explained. Lasers consist of a column of material with a partially reflecting mirror at one end and a totally reflecting mirror at the other. The details of the lasing material need not concern us except that its atomic electrons can be excited to a state from which they find it very difficult to decay spontaneously. The laser is 'primed' by means of an intense flash of light so that a very large number of atoms occupy this excited state. Suppose one of these atoms makes the transition to the ground state and in doing so gives up its energy by emitting a photon.* If it happens to be emitted parallel to the column this photon will make many traversals of the column as it is reflected between the mirrors, which are arranged to be accurately parallel. As it does so this photon perturbs the other atoms inducing them to emit similar photons of the same frequency. The emission process 'avalanches'. Instead of the atoms decaying slowly from their excited state to the ground state they are stimulated to emit photons. All these emitted photons are released in phase with the inducing photon and with nearly exactly the same frequency. The waves which do not travel parallel to the column strike the walls and are absorbed. Those that are parallel are reflected back and forth by the mirrors which are a whole number of wave lengths apart so that the wave keeps in phase. The intensity of the beam builds up rapidly due to all the atoms falling to the ground state and the light eventually emerges from the partially reflecting mirror. It is nearly monochromatic and of high coherence for the reasons outlined above. It is intense because all the excited atoms have been induced to emit photons cooperatively. The variation in the mirror separation must be very much less than one wavelength in a total column length of about 30 cm. The overall result, therefore, is that the Light has been Amplified by the Stimulated Emission of Radiation — the name laser is an acronym derived from the capitalised initials.

* A word we used in an earlier section to describe a short wave train of light.

Optics and the Wave Nature of Light

Most of the applications of laser light stem in a fairly obvious way from its properties. For instance, the extreme directionality of the beam allows it to be used by surveyors to line up equipment in inaccessible places, such as when laying pipelines or underwater tunnels. The tunnels of the new San Francisco Bay rapid transit system were indeed aligned under the Bay, between Oakland and San Francisco, by this means. However the laser is a quite novel device and new applications are continually being thought of; this was not the case with, for example, the transistor which was a replacement for the triode valve and hence had predictable uses.

Another fairly obvious use of the laser is the controlled utilisation of its power. Our description of the mechanism of laser operation suggested that it always leads to sudden bursts of power (pulsed lasers). It is also possible to run lasers continuously, and greater than 10^5 watts of continuous power can thereby be achieved. By focussing such laser beams to a sharp point it is possible to produce even greater power concentrations, of the order of 10^9 watts per square centimetre. (Imagine trying to focus a large conventional source down to a small image. The light from an ordinary source is emitted independently by each atom so that the photons go in random directions at random times. Therefore the energy will be radiated in all directions. Efforts to collect and focus this energy with a lens cannot do better than simply reproduce the power concentration of the light source itself.) On the other hand, the laser can produce a beam of light which can be totally collected by a lens, and due to the coherence of the light it can be focussed down to an image whose size is limited only by diffraction. The result is the tremendous power concentrations just mentioned, and the ability of this concentrated power to vapourise material quite suddenly is the basis of many laser applications. For instance, print may be erased by vapourising it from the page without burning the paper; and precision electrical resistances are made by vapourising incredibly small amounts of material until the desired value is reached. One of the most impressive industrial uses is in drilling tiny holes of awkward shapes in hard or refractory materials. It is even possible to drill square holes through diamonds this way.

An important use of laser power is in surgery. Eye operations on loose retinas (detached retinas) have been performed by 'welding' the retina back into position with a laser beam. Also so-called 'bloodless' surgery can be performed, whereby the laser beam acts not only as a sharp precision knife, but as a self-cauterising one.

The localised heating it produces causes all the cut blood vessels to be instantaneously sealed. At present the main obstacle to the laser's further development as a surgical tool seems to be that of directing the light to the required place without using a complex system of mirrors and prisms which requires a clumsy counterweight system for its easy manipulation.

One of the most interesting applications of lasers is called *holography*. This is a way of recording all the relevant information about an object in such a manner that it is afterwards possible to generate a three-dimensional image of the object from the recorded information. The type of information required for such a 'photograph' is, for example, the set of surface coordinates of the object. If a unique mathematical transformation were done on the set of coordinates then the transformed information could be used to reconstruct the three dimensional image (as it is possible by reading numbers from a table to draw a graph). The inverse transformation would have to be performed in order to get back to the coordinates (just as one could get back the numbers in the table by reading them off the graph). Now the diffraction pattern of light scattered from an object is itself a transform or coded record of the object. Given such a diffraction pattern it is possible to reconstruct a three dimensional image of the object. It is necessary to record all the information in the diffraction pattern, that is both the phase and the amplitude at each point in the pattern. The way in which this is done is to shine a laser beam onto the object and to arrange for the scattered beam to interfere with another laser beam (called the reference beam) thus forming a set of interference fringes on a photographic plate. The contrast (relative brightness) of the fringes is a measure of the amplitude of the diffracted beam. The position of the fringes records the phase of the diffracted beam. Thus the photographic plate is the desired record of the transformed information. The trick in holography is to shine the reference laser beam back onto the developed photographic plate. The beam is diffracted and forms a three dimensional image of the original object. Thus the coding of the information and also its decoding are done by light beams. If the information on the photographic plate were to be used in order to calculate the image of the original object a large amount of digital computation would be required. This trick of reconstructing the image by re-diffracting light from the hologram is effectively analogue computation.

6

MODERN PHYSICS: PARTICLES, WAVES AND PROBABILITIES

The physical phenomena we have studied so far in this book were all well known to, and quite well understood by, the scientists of the late nineteenth century. Between a date which we may put rather arbitrarily at 1890 and the present day, a revolution has occurred in man's understanding of the natural world, comparable to the revolution in outlook that followed the discovery of America by Columbus. In this chapter we shall try to explain why and how this revolution came about and why it is immensely significant for all scientists, not only physicists.

The revolution was primarily one of *scale*. So-called 'classical' physics, of the pre-1890 type, was concerned with events and processes that could be measured in terms of seconds, grams and metres, by human observers using instruments (clocks, balances, measuring rods) that had been devised with human dimension and time-scales in mind. True, microscopes and telescopes had been invented around 1600, enabling biologists and astronomers to make rapid strides in the next two centuries, and it was primarily because of the relatively advanced state of optical instruments that the study of the nature of light, for instance, had extended the distance scale down to the Ångstrom region. But these were only small perturbations to the general rule that it was man, the observer, who determined the scale, and hence the nature, of the phenomena observed. It should be clear now why much of what may loosely be termed 'modern' physics is puzzling to the layman and difficult to understand, even for the scientist trained in biological disciplines. It is because the scale involved is no longer the human one, and consequently the common-sense approach, derived ultimately from our experience, will not work. A whole new set of concepts and rules has to be learnt.

Such a revolution in scale can lead in two directions, upwards and downwards. The former direction is associated with our knowledge of the large-scale structure of the world, the existence, properties and evolution of stars, galaxies and, ultimately, the universe itself. Its progress was hastened by the construction of larger telescopes and more precise measuring instruments. The jumping-off point for this revolution was probably the discovery in 1883 that the velocity of light in a vacuum, c, is independent of how fast the observer or the source is moving. This seemingly paradoxical discovery was used brilliantly by Einstein as the foundation of his theory of relativity, which underlies all modern theories of cosmology. The velocity c is another of those quantities, like Boltzmann's constant k, which are known as *fundamental constants*, and of which we shall meet several more. Their magnitudes determine all the phenomena of the physical world, though why they should have the actual values they do, no one knows. It is the business of the physicist to find this out.

This outward revolution, though fascinating to the astronomer and physicist, does not impinge much on our immediate senses. However, the inward revolution, the study of events that take place in short times and over small distances, has been of immense significance. From a theoretical standpoint it underlies most of the advances that have been made in biochemistry, genetics, neurophysiology and many other areas of biology. From a practical point of view, its discoveries are fundamental to nearly all the instrument and techniques used in everyday laboratory and field studies. There are two basic conclusions of the revolution, and though we shall be discussing them in greater detail later we mention them here so as to prepare the reader for the remainder of this chapter. The first conclusion is that, in the microscopic world, one cannot state with certainty, 'such-and-such an event *will* occur'. That is, the physical world is no longer deterministic but must be understood in terms of probabilities and likelihoods, as when we toss coins. This lack of determinism is not too remote from our experience; it is characteristic also, though for very different reasons, of such phenomena as the weather: but the second conclusion of the revolution *is* unexpected. It is that quantities such as energy and angular momentum, which to our gross senses are not at all restricted in the range of values they can assume, are allowed to have only certain discrete values, the magnitudes of which are fixed by another fundamental constant known as Planck's constant, h. The numerical value of h is very small, so this 'granular' or 'atomic' structure of energy, say, is only apparent when the energies themselves are very small indeed, of the order of those encountered in single atoms. At this level, energy appears only in discrete 'packets' or *quanta* (Greek *quantum*, a packet) and for this reason the sort of physics we shall be discussing is often called *quantum physics*. Just as we began this book with a chapter on how objects move in classical physics, and called the subject-matter *mechanics*, so a trained physicist would begin a discussion of quantum physics with a chapter (or a book!) on the rules which govern the motion of objects in the microscopic world, calling them the rules of *quantum mechanics*. The necessary mathematical approach would obscure our final goal of giving some physical understanding or 'feel' for the subject, so we shall instead approach quantum physics from a much more descriptive viewpoint, first explaining how and why the quantum revolution occurred.

Gas discharges and atomic structure

The ancient feeling that matter consisted of indivisible atoms was codified and placed on a sound experimental basis by Dalton, Proust and other chemists of the early nineteenth century. They were unable to explain, however, *why* there exists only a relatively small number of atomic elements, each with its own characteristic set of properties such as valence. We now know that the main reason for their ignorance was not lack of scientific ability but the fact that the energies required to separate atoms into smaller parts were not available to these men. Chemical experiments involving atomic separation and combination can be done with the aid of heat energy alone, provided by fires and furnaces, but to reveal the structure of atoms it is necessary to apply much greater energies which can only easily be produced by electrical methods. As soon as dynamos and generators were introduced and electricity became fairly well understood, it was possible to apply large voltages to materials. The obvious substances to investigate were gases, since their pressure could be altered by the new vacuum pumps that were being introduced at this time. From 1850 onwards much work was done with partially evacuated glass tubes to which high voltages were applied, and the basic facts discovered, with their modern explanation, were as follows.

At high pressures nothing happens. As the pressure is reduced, the atoms are partially *ionised*, releasing *negative electrons* which travel to the anode and *positive ions* which travel, more slowly because of their smaller mean free path, to the cathode. Thus a small current flows in the tube. At lower pressures the electrons (and ions to some extent) are accelerated by the electric field to such a degree that, before they reach the electrodes, they can collide with and ionise further atoms. This avalanche of ions causes the current passed to rise rapidly and a bright glow of light to appear, of a colour characteristic of the gas in the tube. (The light is due to recombination of ions and electrons to form neutral atoms, and we shall discuss its spectrum in a later section.) At still lower pressures more complicated effects are seen, but if at last the pressure is sufficiently low that the mean free path is comparable with the distance between electrodes, a stream of electrons can be seen (from their glowing track) to move from cathode to anode, and a stream of positive ions travels in the reverse direction.

The properties of the electrons (originally known as *cathode rays* and of the positive ions were extensively investigated by applying various electric and magnetic fields, and by altering the gas filling, its pressure and the electrode material and voltage. The charge and mass of the ions varied from gas to gas, but the properties of the electrons were found always to be the same. Their *electric charge*, denoted by the symbol e, has been measured as $1 \cdot 602 \times 10^{-19}$ coulomb, and their *mass* m_e, as $9 \cdot 11 \times 10^{-31}$ kg. Let us put these figures in perspective with a simple example, repeated from Chapter 4. Approximately 1 amp of current flows through a powerful flashlight bulb, which means that 1 coulomb per second or about 6×10^{18} electrons per second are passing. Since about 10^{27} electrons constitute a mass of 1 gram, the bulb would have to be alight for $\dfrac{10^{27}}{6 \times 10^{18}}$ seconds, or approximately five years, before 1 gram of electrons had flowed through the filament.

The mass of the positive ions, though variable, is always many thousands of times greater than that of the electrons; their charges, however, are found to be always some simple multiple of the electronic charge. The modern idea of atomic structure is therefore the following; the atoms consist of a relatively heavy, positively-charged core or *nucleus*, surrounded by a number of much lighter electrons. The charge of the nucleus is equal to the total charge of the electrons, making the atom electrically neutral. The chemical properties

Modern Physics: Particles, Waves and Probabilities

of the atom are due to the electrons, the outermost of which are relatively easily detached. Their total number is known as the *atomic number*, usually denoted by the symbol Z. Each element has a different value of Z, which ranges in nature from 1 (hydrogen) to 92 (uranium). An analogy is often drawn between the electrons moving around the nucleus and the planets rotating about the sun; though descriptive it is not very apt, as we shall see shortly, except in so far as it suggests the great emptiness of atoms. As mentioned in Chapter 2, atomic diameters are of the order of 10^{-10}m, whereas nuclear diameters are some one hundred thousand times smaller that this. Thus seemingly solid objects are mostly empty space. Their rigidity, opaqueness and other 'solid' properties are due entirely to the electronic structure of the material, the electrons reflecting light and, by virtue of their mutual repulsion, acting as the 'springs' holding the atoms apart.

The photoelectric effect

The first direct observation of a quantum effect (though its significance was not appreciated immediately) was made in 1887 by Hertz, the discoverer of radio waves. He found that the current flowing in a discharge tube was increased if light from an arc-lamp was shone on the cathode, and it was soon shown that this was due to the light itself ejecting electrons from the cathode surface. This phenomenon is known as the *photoelectric effect*, and it was at once shown to have some rather unusual properties. Very careful measurements showed, first, that if the frequency of the light was below some threshold value, *no* photoelectrons at all were emitted, no matter how bright the light, and second, that for frequencies above the threshold value, *some* photoelectrons always appeared no matter how dim the light. Now this behaviour is very strange; according to the last chapter, light consists of wave trains and if these are of very small amplitude (intensity) one would not expect to find sufficient energy gathered into a volume the size of an atom to knock out an electron.

The explanation is this. Although all the properties of transmission, reflection, diffraction and so on can be explained by the wave nature of light, we now believe that light is *emitted* (as in discharge tubes) or *absorbed* (as in the photoelectric effect) in small packets or quanta. The quantum of light is usually called a *photon*. Thus we should consider very dim light falling on a surface not as a wavefront of small amplitude striking the whole area continuously, but as a number of photons which pepper the surface and as the light becomes dimmer, strike fewer and fewer electrons.

What is this photon energy, and how is it related to the frequency If the photoelectric surface is the cathode of a vacuum tube, the ejected electrons will be pulled towards the anode and will acquire energy. Conversely, if the surface is the anode the electrons, even though ejected with some finite energy, will be pulled back towards the anode by the electric field. By measuring the voltage, V, needed *just* to stop the electrons from leaving the surface, one can measure their kinetic energy; it is simply eV, where e is the electron charge and V measures the change in electric potential. It is found that this kinetic energy, $\tfrac{1}{2} m_e v^2$, is given by the equation

$$eV = \tfrac{1}{2} m_e v^2 = h\nu - \phi \tag{6.1}$$

where ν is the frequency of the light, and h and ϕ are two constants The latter, ϕ, depends on the material used, its state of polish, and so on; it measures the energy needed to drag the electron out of the interior across the surface, and is called the *work function* of the surface. The quantity h, however, is found to be a constant, and is in fact Planck's constant referred to earlier. Its value is $6\cdot63 \times 10^{-34}$ J s. Thus all the phenomena of photoelectric emission can be explained if we assume that light photons of frequency ν have energy E given by

$$E = h\nu \tag{6.2}$$

Equation (6.2) is our first example of quantisation.

Example: On this microscopic scale, energies are conveniently measured in *electron volts* (eV). One eV is the energy gained by an electron accelerated through a potential difference of one volt, and is therefore equal to $e \times 1 = 1\cdot6 \times 10^{-19}$ J. Light of wavelength 4000 Å $= 4 \times 10^{-7}$ m therefore consists of photons of energy

$$h\nu = \frac{hc}{\lambda} = \frac{(6\cdot63 \times 10^{-34}) \times (3 \times 10^8)}{4 \times 10^{-7}} \approx 5 \times 10^{-19} \text{ J} \approx 3\cdot1 \text{ eV}.$$ Now most common metals have work functions of a few eV; that of sodium for instance, is $2\cdot3$ eV. Consequently electrons of kinetic energy as large as $(3\cdot1 - 2\cdot3) = 0\cdot8$ eV can be emitted from a clean sodium surface by 4000 Å light. Conversely, the threshold frequency for sodium is given by finding the frequency corresponding to $2\cdot3$ eV; it is

$$\frac{(2\cdot3) \times (1\cdot6 \times 10^{-19})}{6\cdot63 \times 10^{-34}} = 5\cdot6 \times 10^{14} \text{ Hz, corresponding to a wavelength}$$

of $\dfrac{3 \times 10^8}{5\cdot 6 \times 10^{14}}$ = $5\cdot 4 \times 10^{-7}$ m or 5400 Å.

A familiar application of the photoelectric effect is the photographic light-meter, in which light is allowed to strike a metal, often an alloy such as caesium-antimony, which has an extremely low work-function. The ejected electrons can flow as a current which is registered on a meter. The device can be made even more sensitive by amplifying the current, and is then known as a 'photomultiplier' or 'image intensifier'. Even these devices, however, are barely a match, so far as sensitivity and adaptability goes, to the human eye, which in a dark-adapted state can detect a flash of light containing fewer than ten photons. The effect here is of course not strictly photoelectric but photochemical, the retinal pigment (rhodopsin) being bleached by a chemical change following absorption of the light energy of a single photon. However the energy available is still related to the frequency by an equation similar to eqn (6.1), though the 'work function' is far lower than for any photoelectric process, so low in fact that the threshold frequency is well into the infra-red. Consequently, the variation of eye response across the visible spectrum is determined mainly by the absorption (that is, the colour) of the retina, though superimposed on this there is the expected fall-off from blue to red.

Similar arguments apply to the mechanism of photosynthesis in plants, the first step being the absorption of single photons by the chlorophyll contained in small chloroblasts on the leaves. Again the 'work function' is very low.

The production and diffraction of X-rays

In 1897 Roentgen, whilst studying whether light from discharge tubes affected photographic plates in the normal manner (it does), found 'fogging' and evidence of exposure not only on plates that he exposed deliberately but also on unwrapped, light-tight plates left in the laboratory. Evidently some source of penetrating radiation other than visible light was affecting the photographic emulsion, and Roentgen's initial name of *X-rays* for this mysterious effect has stuck. It is now known that X-rays are simply electromagnetic waves of exactly the same nature, and moving at the same velocity, c, as visible light, but of very much shorter wavelength. This short wavelength (high frequency) accounts for their penetrating power, and there is no need here to discuss the medical and technological usefulness of X-rays. Roughly speaking, the production of X-rays in a discharge tube is the converse of the photoelectric effect. In the latter photons eject electrons from a metal surface, and in the former the energetic electrons striking the anode eject photons from it. Equation (6.1) still applies, only now we can determine the frequency of the ejected photons once we know the energy (that is, the accelerating potential in the discharge tube) of the incident electron. Since these potentials are often tens of thousands of volts the work function is very small by comparison and is usually ignored, whilst the corresponding photon frequencies are $\sim$ 10 000 times larger than in the photoelectric effect. As regards terminology, X-rays of high (low) frequency are often called *hard* (*soft*), reflecting the difference in their relative penetrating power. Thus, raising the voltage in the discharge tube will harden the X-rays and so increase their penetration. However, not all the X-rays have the maximum energy (frequency) that is predicted by this argument. The reason is that when the fast electrons strike the anode they are very quickly slowed down by collisions with other electrons and atoms, and either never eject a photon at all, or only do so after their energy has been much reduced. Thus the X-ray frequencies range from a maximum determined by the tube voltage down to very low values indeed. The probability of emitting X-rays which are near the maximum frequency is greatest for materials which have a high atomic number; one also requires the anode to be of a very tough metal to withstand the heating caused by the impact of the intense electron beam. For these reasons anodes of tungsten or molybdenum are commonly used, and a typical modern hard X-ray tube (often called a *Coolidge tube* after its principal developer) will have a hollow, water-cooled tungsten anode and be run at a potential of 100 000 volts or more.

Obviously the most important property of an X-ray is its wavelength. How may this be measured? The wavelength of visible light is most easily and accurately found by using a spectrometer with a diffraction grating, as was described in the last chapter. Now a diffraction grating is simply a set of alternate opaque and transparent strips whose width and separation are of the same order as the wavelength to be measured. What is required is a similar 'diffraction grating' for the much shorter X-rays, and early in this century Bragg suggested that the regular rows of atoms in crystals could be used for this purpose. Atomic sizes and separations, as was seen in Chapter 2, are of the order of several Ångstrom units, corresponding to X-ray energies of several thousand electron volts, and since the atoms are fairly opaque to X-rays the periodic rows of atoms found in crystal lattices make ideal three-dimensional 'diffraction gratings'. Most such devices, in fact, utilize reflection rather than transmission of the X-ray beam, and in this case it is easy to find the relation between the X-ray wavelength and the angles at which constructive (or destructive) interference occurs amongst the reflected waves. Referring to Fig. 6.1, suppose a parallel beam of X-rays is incident at some angle θ (the *glancing angle*; note that θ is not the angle between the incoming beam and the normal to the surface, a departure from optical convention which can confuse students) to the surface of a crystal which, we will assume, has sheets of

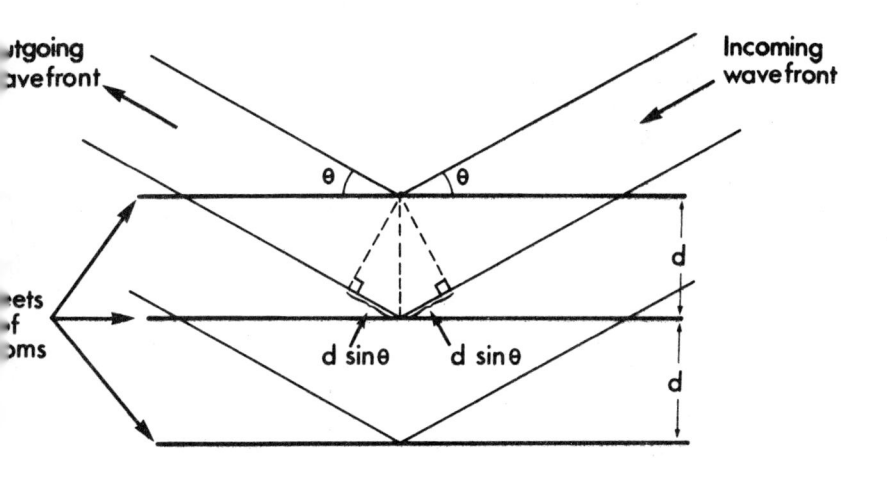

Fig. 6.1
Bragg reflection of X-rays from a crystal. The sheets of atoms from which scattering occurs are shown a distance d apart, and the path difference of $2(d \sin \theta)$ between incoming and outgoing wavefronts is shown.

atoms a distance d apart running parallel to this surface. Scattering from these atoms will take place in all directions (and so the term 'diffraction' is more appropriate than 'reflection' or 'transmission'), but it should be clear from the diagram that a certain path difference, equal to $2d \sin \theta$, is introduced between two portions of a wavefront which scatter at angle θ from two adjacent sheets. If this path difference is exactly equal to one (or more) wavelengths then, on recombining the scattered wave fronts, *constructive interference* will occur. Now since the X-rays pass through, and scatter from, an enormous number of sheets of atoms with very little attenuation, the constructively interfering diffracted wavefront will build up a large amplitude, and can be detected, either on a photographic film or by one of the devices described in the next chapter.

Thus the X-rays appear to be reflected, as if from a mirror, at angles θ such that

$$n\lambda = 2d \sin\theta \qquad (6.3)$$

where n is an integer, this being known as the *Bragg equation*. Various crystals with well-known atomic spacings are universally used in such *Bragg spectrometers* to measure X-ray wave lengths.

The converse use of X-ray diffraction, that is using X-rays of known wavelength to deduce the structure of new or complex crystals, is also very important. If λ is known and θ measured, d can be calculated from eqn (6.3), and by orienting the crystal in various directions the separation of the various planes of the crystal lattice can be found. This, basically, is how the structures of many organic molecules such as haemoglobin, and, more sensationally, DNA ('the double helix') have been found. See the book by Watson, p. 220.

A topical example of a new mineral structure that has been determined in this way is that of *armalcolite*, $(Fe, Mg)Ti_2O_5$. This mineral was discovered amongst the lunar samples brought back to Earth by the Apollo 11 astronauts. (The name is an acronym derived from ARMstrong, ALdrin and COLlins, the crew members.) X-rays from an iron target were used, their wavelength being 1·936 Å. The most prominent interference fringes occurred at angles of 16·2°, 34° and 57° for n = 1, 2 and 3 respectively. These angles correspond to an interatomic spacing of 3·465 Å. Observations of other, less prominent fringes allowed the whole structural pattern of armalcolite to be deduced.

The wave nature of electrons

In 1926 Davisson and Germer, two physicists working at Bell Telephone Laboratories in the USA, had what turned out to be lucky accident. They were conducting a routine study of electrons scattering from a sheet of nickel when a liquid air bottle shattered and the clean surface of their target became heavily oxidised. After lengthy heating, first in a reducing atmosphere and finally in vacuum, they managed to remove the oxide layer, but, on resuming their experiments at the point they had previously reached, were perplexed to find that the scattering pattern had completely changed. Instead of the rather diffuse and featureless reflections that had previously occurred, they found now a series of intense bands or fringes, separated by very faint minima; in other words just the sort of pattern one would expect from X-ray diffraction. Following this up they became convinced that the electrons themselves were in some way managing to produce these interference patterns which, as you will realize from the last chapter, can only be explained in terms of wave motion. Just as, in the photoelectric effect, waves had taken on the characteristics of particles, so now it seemed that particles could sometimes behave like waves!

As it happens, only a few years previously a Frenchman, de Broglie, had written a remarkable doctoral thesis in which he predicted, on rather tenuous theoretical grounds, that this phenomenon should occur. Moreover, he had even suggested that the 'wavelength' λ of these electrons was given by the equation

$$p = \frac{h}{\lambda} \qquad (6.4)$$

where p is the momentum of the electron and h is the same Planck's constant that we have met before. The initial reaction to de Broglie's 'waves' (which are termed *matter waves*) was one of scepticism — how could a particle exhibit such properties? A series of experiments by Davisson, Germer and others soon demonstrated, however, that the wavelength of electrons of a given momentum, as measured from the diffraction patterns using the Bragg equation, was indeed exactly that predicted by eqn (6.4). Thus, just as electromagnetic waves can on occasion exhibit all the properties of localized particles, so conversely particles can exhibit wave-like properties. This possession, by the same physical object, of two apparently complementary sets of properties is known as *duality*.

What is the nature of these matter waves? It was at first suggested by de Broglie that an electron should be thought of, not as a tiny hard sphere, say, but as a small 'packet' consisting of a number of waves, and any measurement of where the electron is, or what its momentum is, really measures roughly where the wave packet is, and what its corresponding momentum is, as given by eqn (6.4). Although we shall modify this interpretation in a later section, its usefulness now is that it leads to a fundamental principle of quantum theory which is important in all that follows. This is the *uncertainty principle* and we can easily derive it for the matter waves.

Suppose there are m wavelengths in the wave packet representing the electron. They don't all have the same amplitude; rather, this is greatest somewhere near the centre of the packet and decreases to zero at the ends, and very close to the ends it will obviously be difficult to say precisely where the wave packet begins. Consequently the number m will be *uncertain* to within, let us say, about ± 1 wavelength. Now let us see how precisely we can measure the position (call it x) and the momentum, p, of the electron. The uncertainty in x, that is Δx, is about the total length of the packet, $m\lambda$, since we know only that a position measurement will give a result somewhere in the packet. The momentum is given by eqn (6.4), so an uncertainty Δp, corresponding to a wavelength uncertainty $\Delta \lambda$, is given by

$$p \pm \Delta p = \frac{h}{\lambda \mp \Delta \lambda}$$

$$= \frac{h}{\lambda \left(1 \mp \frac{\Delta \lambda}{\lambda}\right)}$$

$$\approx \frac{h}{\lambda}\left(1 \pm \frac{\Delta \lambda}{\lambda}\right)$$

if $\frac{\Delta \lambda}{\lambda}$ is small.* But since the number of wavelengths is uncertain to one part in m, we know the wavelength itself only to this accuracy, that is, $\frac{\Delta \lambda}{\lambda} = \frac{1}{m}$ and so

$$p \pm \Delta p \approx \frac{h}{\lambda}\left(1 \pm \frac{1}{m}\right)$$

yielding $\Delta p \approx \frac{h}{m\lambda}$ (by subtraction of eqn 6.4).

Since $\Delta x \approx m\lambda$ we have finally

---

* The last step in the argument uses the mathematical fact that

$$\frac{1}{1+x} \approx 1 - x$$

if x is small $\left(\text{e.g. } \frac{1}{1 + 0.001} \approx 0.999\right)$.

$$\Delta p \cdot \Delta x \approx \left(\frac{h}{m\lambda}\right)(m\lambda)$$

$$\approx h \tag{6.5}$$

This is one form* of the uncertainty principle. It states that we *cannot* measure both the momentum and the position of a particle, at a given instant, to an accuracy better than that given by eqn (6.5). The word 'cannot' is to be understood in an absolute sense; it is not that our instruments are too poor in quality, it is rather as a result of the nature of particles themselves. One way of seeing this is to realize that any attempt to measure the position of an object, by bouncing light off it, say, is bound to disturb the object to some extent, and hence change its momentum.

The uncertainties involved are staggeringly small in everyday life. Suppose a microscope has a resolution, that is a Δx, of 0·1 μm. Then the corresponding Δp is about 10^{-26} kg m s^{-1}; we can divide Δp by the mass to find the uncertainty in our knowledge of the velocity, and for a typical small object such as an erythrocyte this turns out to be about 10^{-4} μm s^{-1}, far below the resolution limit. The principle only reveals its true power when applied to systems of atomic dimensions. Suppose an X-ray diffraction experiment reveals that the atoms of a crystal are spaced, on average, about 5 Å apart. This can be taken as Δx; calculating as before, and inserting typical atomic masses, we find a velocity uncertainty of about 10^2 m s^{-1}, corresponding to an energy of about 10^{-2} eV. This represents the amount of energy that must be possessed by an atom whose position is so accurately known (and incidentally shows that, even at the absolute zero of temperature, atoms are never completely at rest).

Besides these theoretical consequences of the wave nature of particles, there are much more immediate practical applications, the most noteworthy probably being the *electron microscope*. In this instrument a beam of electrons is passed through (or scattered by) an object, and then collected onto a sensitive surface such as a photographic plate or a TV screen by a system of electric and magnetic 'lenses' much as a light beam is focussed by glass lenses in a

---

* The uncertainty principle can be written in other forms, in terms of quantities other than momentum and position. For instance, attempts to measure energy E at a particular instant t have an inescapable inaccuracy given by

$$\Delta E \cdot \Delta t \approx h$$

conventional optical microscope. The great advantage of the electron instrument is its much higher resolution. Remember that the resolving power is roughly equal to wavelength divided by numerical aperture; clearly the resolving power of optical microscopes is limited by the wavelength, say 5000 Å, of the light used, but the matter waves of even moderately-energetic electrons have far shorter wavelengths than this. From eqn (6.4)

$$\lambda = \frac{h}{p} = \frac{h}{\sqrt{2mE}}$$

where E is the electron energy. Thus an electron of 10 000 electron volts has a wavelength

$$\lambda = \frac{6 \cdot 6 \times 10^{-34} \text{ Js}}{\sqrt{(2 \times 9 \cdot 11 \times 10^{-31} \text{ kg}) \times (10\,000 \times 1 \cdot 6 \times 10^{-19} \text{ J})}} = 10^{-11} \text{ m or } 0 \cdot 1 \text{ Å}.$$

Thus the resolution, for the same numerical aperture, is 50 000 times better than that of an optical microscope.*

Of course there are disadvantages in the electron microscope technique. Samples can only be examined in vacuum, necessary for the electron beam; the intense beam easily damages the samples; the latter must be very thin (several microns only) to let the beam through; and so on. But the enormous increase in resolution often makes the technique the only available one for examining very fine structure. In the bibliography we cite a recent article dealing with advances in electron microscopy.

Quantization and atomic spectra

Besides the duality between waves and particles which we have just been discussing, the other great discovery of quantum physics is that the energy possessed by atomic systems cannot have any value whatsoever but is constrained by certain quantum rules to take only various discrete values known as *energy levels*. We shall now show how these rules follow very easily from the ideas of the last section, and what their consequences are.

Recall the principles of atomic structure in an earlier section, where we stated that the central heavy nucleus of the atom, with

---

* This resolution is predicted also from the uncertainty principle. When the electrons are scattered by the object, the maximum uncertainty in their final momentum, Δp, can only be as great as the momentum, p, itself (and then only if the numerical aperture is unity). Consequently the uncertainty in object position, Δx, is equal to $\frac{h}{p}$, which is equivalent to the expression above.

Modern Physics: Particles, Waves and Probabilities

positive charge Ze, is surrounded by a cluster of Z electrons, each of charge −e. We said then that these electrons are often considered to be circling round the nucleus; why should we suppose this? The answer is that if they were not they would collapse into the nucleus by virtue of their mutual electrostatic attraction, just as, if the Earth were to cease rotating round the sun, it would fall into it because of the gravitational attraction. The electric (or gravitational) force is always present and causes an acceleration, but if the electron (or the Earth) is circling round the nucleus (sun) this acceleration can be just equal to the inward acceleration due to this circular motion (see Chapter 1).

For simplicity consider an atom such as hydrogen or singly ionized helium, which has only one electron circling round a nucleus of charge Ze. Then, as was shown in Chapter 1 (p. 24), the radius of the orbit, r, and the kinetic energy of the electron, T, are related by the expression

$$r = \frac{Ze^2}{8\pi \epsilon_0 T} \quad (6.6)$$

According to our ideas of duality this circling electron may be likened to a wave packet, which evidently 'turns back on itself' around the orbit. Thus we will expect to get interference effects of the wave with itself, as discussed in Chapter 5, and the interference will be destructive *unless* there happens to be *exactly* a whole number of wavelengths around the orbit circumference. In this case the interference will be constructive and we have so called *standing waves*, with radii given by

$$2\pi r = n\lambda$$
$$= nh/p = nh/\sqrt{2mT} \quad (6.7)$$

Only orbits whose radii satisfy eqn (6.7) will be allowed; we can now substitute for r in eqn (6.6) to obtain

$$\frac{nh}{2\pi\sqrt{2mT}} = \frac{Ze^2}{8\pi \epsilon_0 T}$$

and so $T = \dfrac{mZ^2 e^4}{8 \epsilon_0^2 h^2 n^2}$

We now state without proof that the potential energy of the electron is negative and equal to twice its kinetic energy; hence the *total* energy E of the system is $T - 2T = -T$, and so

$$E = -\frac{mZ^2e^4}{8\epsilon_0^2 h^2 n^2} \tag{6.8}$$

$$= -Ry\left(\frac{Z^2}{n^2}\right)$$

where Ry is a constant. This is the first quantum rule of atomic structure, stating that the energy of successive allowed levels is inversely proportional to the square of the integer n, which is called the *principal quantum number*. The constant Ry is the *Rydberg constant* and has the value $2 \cdot 18 \times 10^{-18}$ J (or $13 \cdot 6$ eV).

Since only certain energy levels are allowed an atom can only absorb (or emit) energy equal to the difference in energy between two levels; each such *quantum jump* will therefore result in one quantum of radiation, or photon, of a particular energy (wavelength) being absorbed or emitted. This accounts very well indeed for the observed emission lines in the spectrum of hydrogen, and the corresponding absorption lines seen in the sun's spectrum.*

Jumps to (or from) the level with $n = 1$ constitute one series of lines, jumps to or from $n = 2$ another series, and so on. These series are known as Lyman, Balmer, etc., after their discoverers.

Example: Calculate the wavelengths of the first two lines in the Lyman and Balmer series. The difference in energy between two quantum levels with principal quantum numbers m and n is

$$\Delta E = Ry\left(\frac{1}{m^2} - \frac{1}{n^2}\right)$$

For the Lyman series $m = 1$, and for jumps to the levels with $n = 2$ and 3 respectively, we have

$$\Delta E_1 = 13 \cdot 6 \left(1 - \frac{1}{4}\right) \text{eV}$$

$$= 10 \cdot 2 \text{ eV}$$

and $\Delta E_2 = 13 \cdot 6 \left(1 - \frac{1}{9}\right)$ eV

$$= 12 \cdot 1 \text{ eV}$$

corresponding according to eqn (6.2) to wavelengths of 1216 Å and 1026 Å respectively. Note that these are in the far ultra-violet. For the Balmer series $m = 2$, and for jumps to the levels with $n = 3$

---

* These so-called Fraunhofer lines are due to the relatively cool layers in the sun's upper atmosphere absorbing radiation emitted by the hot layers beneath. Since over 90% of the sun consists of hydrogen, the hydrogen absorption lines are naturally the most prominent.

Modern Physics: Particles, Waves and Probabilities

and 4, we have

$$\Delta E_1 = 13 \cdot 6 \left(\frac{1}{4} - \frac{1}{9}\right) \text{eV}$$

$$= 1 \cdot 89 \text{ eV}$$

and $\Delta E_2 = 13 \cdot 6 \left(\frac{1}{4} - \frac{1}{16}\right) \text{eV}$

$$= 2 \cdot 55 \text{ eV}$$

corresponding to wavelengths of 6573 Å and 4869 Å. These lines are in the visible region and indeed the Balmer series was the first to be found.

Evidently the photon wavelengths depend on the values of m and n, and also of course on Z. If Z is large enough very short wavelength photons may be emitted, energetic enough to be called X-rays, and these are in fact produced by many X-ray tubes. They appear as sharp lines on top of the broad spectrum that has already been described, and their wavelengths are characteristic of the anode element used, e.g. molybdenum and tungsten give characteristic X-rays at quite different frequencies. Of course the accelerating voltage used must be sufficiently high to cause the relevant quantum jump.

Equation (6.8) indicates that the energy of bound electrons (their *binding energy*) is a negative quantity, so if they are given sufficient energy to raise them to zero total energy or above, they will become unbound. (The energy levels that then apply are due to the effect of all the surrounding material, and are so closely-spaced that the electron is effectively free to take any energy at all.) The atom has then been *ionised* and the necessary *ionization energy* is simply

$E_{\text{ion}} = Ry\, Z^2 \left(\frac{1}{n^2} - \frac{1}{\infty^2}\right)$ since the final value of n (the so-called *series limit*) is infinitely large. Most atoms (in a terrestial environment, anyway) are in their *ground states* with $n = 1$, so the ionization energy is just the numerical value of the Rydberg constant times Z^2, i.e. 13·6 eV for hydrogen, and so on.

All these various aspects of atomic spectra show up very nicely in an experiment first performed by Franck and Hertz (1914), and bearing their name. Various voltages are applied across a discharge tube and the current flowing in the tube is measured. In addition to the expected increase of current with voltage, the current/voltage graph shows sudden dips, corresponding to the accelerated electrons gaining just enough energy to raise the gas atoms to the 1st, 2nd, 3 etc. excited state above the ground state; the electron consequently never reaches the anode and the current falls suddenly. As the atom decays back to the ground state it emits a photon. If these photon wavelengths are measured, and the corresponding energies compared with the energy the electrons had (deduced from the voltages at which dips occur), excellent agreement is found.

What are matter waves?

Our interpretation of matter waves has been very naive, despite its predictive successes. Max Born, a German physicist,[*] recognized this in 1926 and suggested that, since it is difficult to conceive of any sort of physical disturbance constituting the waves (as electromagnetic vibrations constitute light waves), they should be thought of as a purely mathematical construction whose physical interpretation is that they give the *probability* of finding the particle at a particular point in space. More precisely, since the wave amplitude is a quantity alternately positive and negative, one should square it to obtain the *intensity* (a positive quantity, of course) which is taken as the probability. This notion that particles such as electrons become somewhat diffused when we try to probe their position too closely is clearly related to the uncertainty principle. We cannot stress too strongly that virtually the whole of modern physics depends on the concept that particle positions cannot be localized, and their momenta *simultaneously* known to an arbitrarily high degree of accuracy. The use of waves to represent particles is just one rather convenient way of expressing this concept in a usable form; there are other ways but we shall not discuss them.

De Broglie's matter waves were rather arbitrarily introduced. In 1926 Schrodinger showed that it is possible to write down an equation whose solution, $\psi(x)$, is the wave amplitude for a particular problem (in fact any solution of Schrodinger's equation has certain undetermined constants in it, and one solves the problem he is interested in by giving these constants particular values). The equation is called Schrodinger's *wave equation*; the solutions $\psi(x)$ are called *wave functions*; and $|\psi(x)|^2$, as we have said, represents the probability of finding the particle at a certain place with coordinate x. (We use just one dimension for simplicity; a real problem will

---

[*] Many of the physicists mentioned in this chapter, Born among them, won Nobel prizes. Quantum physics in the first half of this century had the prestige and intellectual vigour that molecular biology, say, has now.

Modern Physics: Particles, Waves and Probabilities

usually have wave functions depending on x, y, z and possibly t also if they change as time goes on.)

Example: When Schrodinger's equation is solved for the hydrogen atom, the relevant coordinate is the radius, r. The wave functions for the ground and the first excited states are:

ground state $(n = 1)$ $\psi_1(r) = 1{\cdot}48 \; e^{-r/0{\cdot}527}$

first excited state $(n = 2)$ $\psi_2(r) = 1{\cdot}85 \left(1 - \dfrac{r}{1{\cdot}054}\right) e^{-r/1{\cdot}054}$

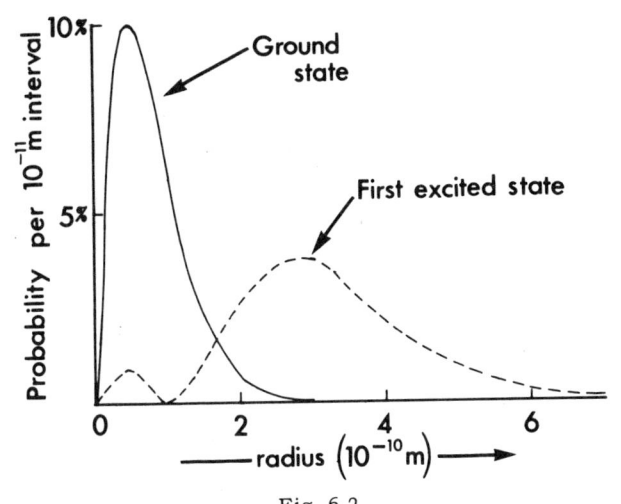

Fig. 6.2
Electron density distributions for the hydrogen atom. We plot the probability that the electron is to be found at some radius r, as a function of r, for the ground state and for the first excited state of the atom.

where r is measured in Ångstrom units ($= 10^{-10}$ m). Hence, by squaring, we can obtain the probabilities *per unit volume of space*. If we wish to find how often a given r value occurs, independent of the direction of the radius vector, we must add up $|\psi(r)|^2$ over the whole surface of the sphere for that value of r. This gives a factor proportional to r^2, since the surface of a sphere equals $4\pi r^2$. The resultant probabilities are plotted in Fig. 6.2. Note that the most probable radius in the ground state is 0·527 Å, exactly the answer we get by the more elementary approach using standing waves (substitute into eqn 6.7 to show this is so).

Further quantum numbers

The simple theory on pp. 160–4 (known as the Bohr theory after the Danish physicist who first developed it) explains very well the spectral lines from hydrogen and other one-electron atoms, but break down if applied to more complex atoms and molecules. To calculate their spectra exactly, and to describe atomic structure and chemical bonding we need to introduce more quantum rules. It turns out that three more rules are sufficient, each of which is associated with another quantum number, just as the rule of quantisation of energy led to the principal quantum number n.

The first rule* is that *angular momentum* is quantised. This is not too hard to understand if we look carefully at the 'dimensions' of Planck's constant, that is the sort of units in which it is expressed. So far we have written h as so many joule-seconds; if we write it in terms of the fundamental quantities mass (M), length (L) and time (T) we see it has dimensions $[ML^2T^{-2}][T] = [ML^2T^{-1}]$, which are just those of angular momentum. (For mass m, velocity v, radius r, angular momentum is mvr which has dimensions $[M][LT^{-1}][L] = [ML^2T^{-1}]$.) Since h is so fundamental to quantum theory it is reasonable to suppose that angular momentum takes only discrete values. It is not simple multiples of h that are allowed, but rather multiples of $\frac{h}{2\pi}$, the 2π arising from numerical factors in Schrodinger's equation; $\frac{h}{2\pi}$ is often abbreviated to $\hbar$ (say 'h-cross'). The rule therefore is

$$\text{angular momentum} = l\hbar \quad (6.9)$$

where l, the *orbital quantum number*, is an integer and can have any value between 0 and $n-1$. Thus if $n = 3$ (the second excited state, since $n = 1$ is the ground state) l can be 0, 1 or 2.

One might ask whether these different values of angular momentum imply different energies for the atom with a particular value of n? The answer is yes, for all atoms except hydrogen-like ones with single electrons; for these there is virtually no energy difference for different l-values, which is why the simple Bohr theory works so well for them. For all other atoms, however, there *is* a difference (often rather small) between the energies, and consequently transitions from one n-value to another give rise to slightly different

---

* All these rules, and the quantum numbers they predict, are derivable mathematically from Schrodinger's equation. However we shall just state them here without proof.

wavelengths depending on the *l*- values of the states. Spectroscopists originally named the different series of spectral lines that resulted the Sharp, Principal, Diffuse etc. from their visual appearance, and these initial letters have stuck, so that states with $l = 0, 1, 2, 3$ etc. are known as *S, P, D, F* etc. (through the alphabet thereafter) states.

The electron orbits in different angular momentum states have different shapes. Only the *S*-states are spherical; *P*-states are dumbell-shaped, with two lobes in which the wave function is large separated by a region in which it is small. *D*-states are more complex still, with several lobes at right angles. The lobes correspond to regions where the probability of finding the electron is large, and so the angular momentum state of the valence electron largely determines the directions in which the bonding force will be stronger.*

We shall mention the other two quantum rules only briefly. First, many spectral lines are found to be, not single lines but double, triple or even more complex. This can be explained by assuming that electrons possess a property called *spin* which is analogous to angular momentum and is similarly quantized, only in units of $\frac{\hbar}{2}$ rather than $\hbar$. An electron can have only one such unit of spin, and the total energy of the atom depends on the direction in which the spin axis of the electron is pointing, relative to the orbital angular momentum. The quantum number associated with the spin of an electron can have only two values, $+\frac{1}{2}$ or $-\frac{1}{2}$ (in units of $\hbar$). The other quantum rule has to do with the direction in which the total (orbital plus spin) angular momentum vector of the atom points. It states that only certain directions are allowed (obviously one must define a standard direction from outside, for instance by applying electric or magnetic fields to the atom, otherwise the direction we are talking about would be arbitrary). The associated quantum number is called m and can have any integral value between $+l$ and $-l$.

If an external magnetic field *is* applied, then the states with different values of s will have different energies, and transitions can occur between them. The energy involved is very small so the corresponding wavelength is large, and the electromagnetic waves are not in the visible region nor even the infra-red, but are more akin to radio-waves. They are known as microwaves and a very important

---

* The terminology used by chemists differs slightly from that of the physicists. The various angular momentum states are termed *orbitals*; thus S-orbital, P-orbital etc. Shared electrons in these states constitute molecular bonds which are given lower-case Greek symbols, thus σ-bond, π-bond etc.

analytical technique known as *electron spin resonance* (e.s.r.) is based on their use. By applying a magnetic field and noting the microwave frequencies which are preferentially absorbed by a material, a very clear idea of its electronic structure can be got.

We thus end up with four quantum numbers, n, l, s and m. One further quantum rule now applies; it is called the Pauli *exclusion principle* and states that *no more than one electron can have a particular set of quantum numbers*. It is this rule which explains the puzzling way in which electrons in atoms occupy successively larger closed shells, resulting in the regularities of the periodic table. For example, consider a P- state with $l = 1$. There are three possible m-values (-1, 0, $+1$) for each of which s can be $+\frac{1}{2}$ or $-\frac{1}{2}$, giving six states altogether. Similarly an S-state has room only for two electrons, a D-state for 10, and so on. Now we can go through the various n-values in turn and determine how many electrons can be accommodated. For $n = 1$, $l = 0$ only, so two electrons is the limit. For $n = 2$, $l = 0$ or 1, allowing $2 + 6 = 8$ electrons. For $n = 3$, $l = 0$, 1 or 2 and $2 + 6 + 10 = 18$ electrons are allowed. If, therefore, an atom happens to have just 2, or $2 + 8 = 10$, or $2 + 8 + 18 = 28$, etc., electrons, they will occupy a series of these so-called 'closed shells'; these atoms are the noble gases helium, neon, argon etc.

The solid state

We now leave single atoms and discuss, rather sketchily since it is a complex subject, how the quantum theory can be applied to atoms which have aggregated together in clusters — that is, to solid, crystalline matter. (We shall not mention liquids, or amorphous solids. The characteristic of crystals which makes them relatively easy to study is their regular lattice-work arrangement of atoms; most materials, including many biological ones, are in fact crystalline when examined closely, and those which are not present very difficult problems to the physicist.)

The concept of discrete energy levels can be applied in a slightly modified form to solids also, and we shall indicate rather schematically how this is done. A single atom has a series of energy levels, each filled with a certain number of electrons, no two of which have the same quantum numbers. Imagine a second identical atom to be brought up very close, so that the electron wave functions overlap. At first sight one would then have pairs of electrons, one from each atom, with identical quantum numbers. The exclusion principle cannot allow this, and what in fact happens is that each of the original levels splits in two, so that there are now twice as many levels which can therefore accommodate all the electrons without any violation of the rules. This is the situation, for instance, in a molecule such as H_2 or O_2. The energy splitting of a particular level is generally small, much less than that between two different levels.

We can repeat the argument for a third atom, then a fourth, and so on, each additional atom adding one more to the *band* of closely-spaced energy levels. Finally, when one has a complete crystal containing say 10^{20} atoms, the electrons will occupy the $\sim 10^{20}$ levels in each of several bands, each band representing what was originally a single energy level. If all the separate atoms were in their ground states (that is with their electrons all in their lowest possible energy levels) one would have a sequence of filled bands, separated by fairly large gaps, overlain by a similar sequence of empty bands. At any temperature other than absolute zero, however, some electrons will be in an excited state (just due to thermal agitation) and these will occupy a few levels in the lowest 'empty' band. The topmost filled band is called the *valence band* and the lowest 'empty' band the *conduction band*, since those electrons it does contain can move easily from one level to another (remember how close the individual levels are) and thus throughout the crystal. These bands are separated by *forbidden gaps*.

The distinction between a *conductor* and an *insulator* is that the latter has a large forbidden gap, and consequently very few electrons manage to reach the conduction band. In diamond, for instance, the gap is 7·2 eV wide; in terms of thermal energy this corresponds to a temperature of about 84 000°K (found by equating kT to the energy gap) so obviously diamond is an exceptionally good insulator at all reasonable temperatures. On the other hand the electronic structure of metallic elements is such that the bands merge to a great extent, and so there is always a partially occupied topmost band in which the electrons are free to move. Such a material will be a conductor. Between these two extremes lie elements such as silicon and germanium, which as crystals have a rather small forbidden gap (about $\frac{3}{4}$ eV for germanium). Since, at room temperature, there will always be a small number of electrons which have been jostled up into the conduction band, these elements are termed *semiconductors*.

If, therefore, a semiconductor is heated its conductivity will increase rather dramatically; this is termed *intrinsic semiconduction* since it is a property of the pure material. Much more important however is so-called *impurity semiconduction*, a property which can be possessed by many materials, amongst them some of biological importance. We shall first explain what the phenomenon is, then discuss some of its applications.

Germanium has four valence electrons. Suppose that we add a small amount of a pentavalent impurity, such as antimony or phosphorus, to a germanium crystal. Two things will happen. First, the crystal lattice will be disturbed by the impurity, the result being in fact that some extra allowed energy levels, associated with the impurities, appear in the forbidden gap, just below the conduction band. Second, each impurity atom finds itself with an extra electron, having five valence electrons compared with the four of its neighbours, and this electron is *donated* to the conduction band. Consequently the introduction of an impurity has increased the conductivity of the material. A very similar thing happens if a trivalent impurity is introduced; now the impurity energy levels become close to the valence band, from which an electron is *accepted* by the impurity atoms to bring it up to equality with its neighbours. The resulting 'hole' in the valence band can be filled by an electron moving into it, and can in effect travel freely through the material just as a gap in a stream of traffic can move rapidly down the road.

We thus get two types of impurity semiconductors, known as *n*-type and *p*-type, in which the current is carried by negative electrons and by so-called 'positive holes' respectively. In effect, one can imagine the *n*-type having a large number of mobile negative charges, and the *p*-type similarly having mobile positive charges. Virtually the whole of modern electronics is based on the properties of these semiconductor materials, which can be prepared with accurately known properties by adding small carefully controlled amounts of an impurity to the pure substance. Consider, for instance, a thin wafer of *n*-type stuck onto a similar *p*-type wafer; if a positive voltage is applied to the *p*-type side and a negative voltage to the *n*-type side, electrons will flow from the *n*-type, through the external circuit and into the *p*-type, there neutralizing the positive holes. Thus a current flows. If the voltages are reversed, however, there are no free electrons on the *p*-type side which are available, so no current can flow. Thus such an *n-p* combination acts as a *rectifier*, i.e. it allows current to pass in one direction only, and performs a job which previously had to be done with a diode vacuum valve. Similarly, two such *junction rectifiers* stuck back-to-back can replace the triode amplifying valve; this combination is known as a transistor.

The advantages of such components, known as *solid-state devices*, over the older vacuum valves are those of cheapness, robustness, reliability, small power requirements, negligible heat output and, above all, small size. Such is the progress of miniaturization that now literally thousands of these components, formed into complex circuits, can be produced in volumes of a few cubic millimetres, by appropriate impurity treatment of the pure substance. Without this sort of technological advance high-speed computers, for example, would be impossible, not to mention cheap radios!

Finally, we reiterate that many biological materials are quasi-crystalline in form, that is consist of a regular lattice of complex molecules. There is evidence that some possess the sort of band structure we have been discussing and should be classed as semi-conductors. A very useful article discussing the biological occurrence of semiconduction is cited in the bibliography.

7

NUCLEAR RADIATIONS

In an embarrassingly literal sense the study of nuclei, the central cores of atoms, has grown explosively in the past half century. The reason for this late development is that the forces between the constituents of a nucleus, the *nucleons*, are so strong that the nucleus is (usually) an extremely stable object, requiring considerable energy to disrupt it. Until the recent development of machines for accelerating particles to high energies the nucleus and its properties remained comparatively unknown. Conversely, the release, in a controlled or uncontrolled manner, of even a very small fraction of the binding energy of such a tightly bound system is a potent source of power. The military, social and economic consequences of such energy release are too well-known for comment.

However, research into ways in which useful energy can be extracted from the strong nuclear force is only one aspect of nuclear physics. The physicist is also interested in the light such studies can shed on the fundamental properties of matter; this is a difficult and specialized field and we shall touch on it only briefly here. Another major field is concerned with the radiations emitted by nuclei and the effects of such radiations on living things. It is a field to which physicists, chemists, biologists and physicians have contributed. The subject is indeed so interdisciplinary that the name 'health physics' has been coined to cover all aspects of the biological effects of radiation. The aim of this chapter is to give the reader a basic understanding of how these radiations are produced, what their effects are, and how these effects can be detected and measured. To do this it is first necessary to discuss what goes on in the nucleus of an atom. (The whole subject is discussed in great detail in the book by Cember cited on p. 220.)

The atomic nucleus

As explained in the last chapter, atoms consist of a small heavy* positively charged nucleus surrounded by a cloud of light, negative charged electrons. These electrons occupy certain energy levels which determine the chemical character of the atom, and there are just sufficient of them exactly to neutralize the nuclear charge. The atomic number Z is the total number of electrons, and therefore also the total number of positive charges in the nucleus. Now it has been conclusively shown that the positive nuclear charge is carried by particles called *protons*, which have a mass some 1836 times that of an electron. This inequality in masses is so great that in many applications one can neglect the mass of the electrons compared with the protons, and consider the whole mass of the atom to be concentrated in the nucleus. However, it is evident that there must be more particles in the nucleus than protons alone. The experimental values of atomic masses turn out to be at least twice as great as one would expect if the nucleus contained only Z protons. It is now accepted that there exist particles of approximately the same mass as protons, but with no electric charge, and that these make up the remainder of the mass of the nucleus. These *neutrons* were first observed in 1932. They are slightly heavier than protons, having a mass about 1839 times that of an electron. Though neutral they interact very strongly with protons through the so-called 'strong nuclear force' which at the very short distances found in the nucleus tends to bind all nucleons[†] very tightly together. This is obviously necessary to stabilise the nucleus, since otherwise the mutual electrical repulsion of the positive protons would fling them apart.

Thus nuclei are made up of neutrons and protons in roughly equal numbers. If we let A be the number of nucleons, the number of neutrons is simply $N = A - Z$. One may calculate the mass of an atom by adding the mass of Z protons to that of N neutrons, and finally adding the mass of Z electrons. Since the electrons' contribution is small the answer should be almost exactly A times the average mass of a nucleon. (Most tables of atomic masses quote masses in terms of the *atomic mass unit*, or *amu*, defined as follows. The mass of a carbon atom, with $Z = N = 6$, is taken to be exactly twelve times some average mass called the amu. A proton, for instance, has a

---

[*] The density of nuclei is phenomenally high — about 10^9 tonnes per cubic centimetre! Taken together with the virtual emptiness of the rest of the atom this results in typical densities of one to ten grams per cubic centimetre for matter in bulk.

[†] Generic name for neutrons and protons.

mass of 1·0073 amu.) If one now compares these expected masses with those found experimentally, one at once finds that the latter are often very far from being integral multiples of the amu. There are two reasons for this. The first is the fact that nuclei of the same chemical element can often exist with different numbers of neutrons, and thus with different masses. Such nuclei are known as *isotopes* and since chemical properties are determined by the electronic structure, the various isotopes are chemically identical. An example is hydrogen, the simplest element with a nucleus consisting of a single proton; it has an isotope, deuterium, whose nucleus contains an additional neutron and which constitutes about one part in six thousand of naturally-occurring hydrogen. Compounds containing deuterium are denser than their analogues containing hydrogen (hence the name 'heavy water'). Some elements have many isotopes (tin has over a dozen!) and since the quoted mass is usually that of the naturally occurring mixture of isotopes, it can often fall between integral multiples of an amu. For example, chlorine occurs as Cl^{35} and Cl^{37}, the superscript denoting the value of A. The isotopes occur in the ratio 3 : 1 and the resulting atomic weight is 35·5. This reason for mass discrepancies is, of course, just due to our considering a large number of particles at once.

The second reason for these mass discrepancies is more subtle and underlies our introductory remarks about the energy stored in nuclei. In Chapter 1 (p. 25) we stated that Einstein summed up his discovery that mass and energy are, in a sense, interchangeable, by the equation $E = mc^2$. For many years this equation was only a theoretical curiosity, since energies large enough to test it were unattainable. Now that this is no longer so, and accelerators can produce particles of enormous energies, it has been completely vindicated. Its relevance here is this; when nucleons bind together to form a nucleus, an extremely large amount of energy is released. This energy is provided by a slight reduction in the sum total of masses present, that is to say, the resultant nucleus is less massive than its constituent nucleons. This is the second reason for non-integral values of atomic weight.

There are various ways of describing this reduction in mass of complex nuclei; two useful ones are the *mass defect* (the difference between the nuclear mass and that of the constituent nucleons) and

Fig. 7.1

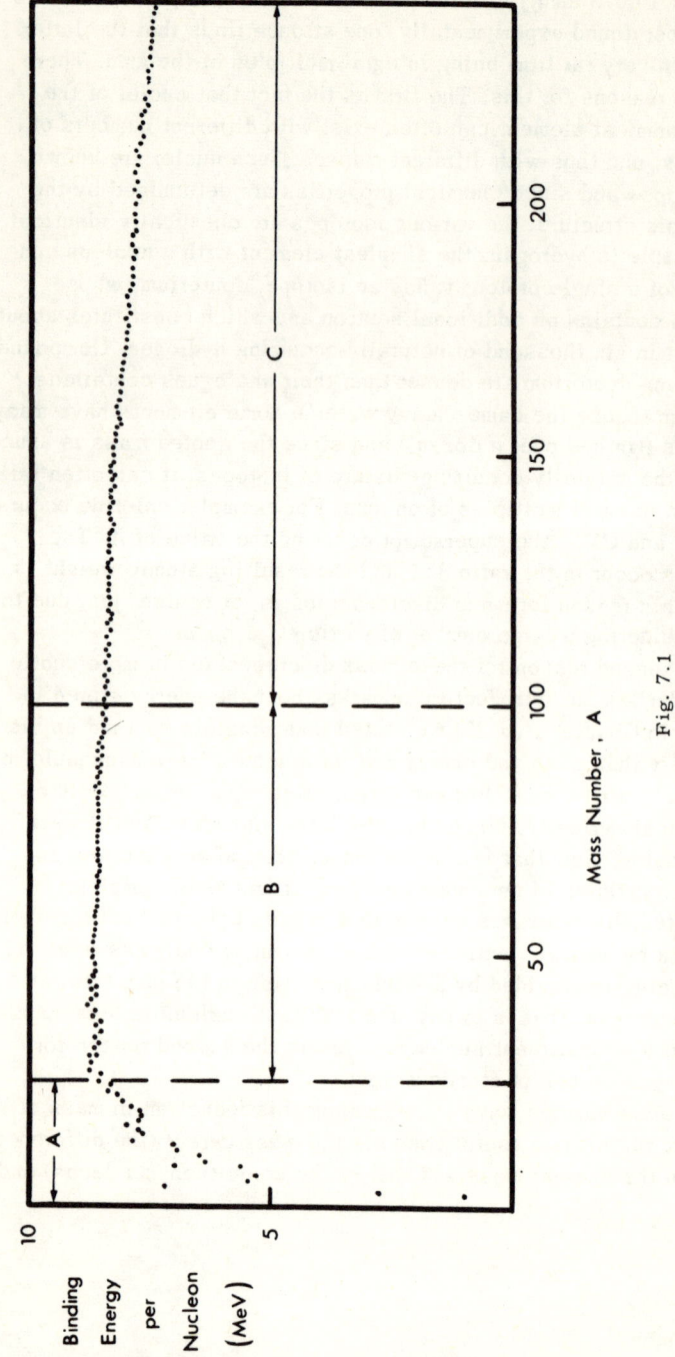

Plot of binding energy per nucleon versus mass number A. The approximate boundaries of the three regions A, B, C referred to in the text are indicated.

Nuclear Radiations

the *binding energy** (the same quantity, expressed in terms of energy using the expression $E = mc^2$). The energy equivalent to 1 amu is 931·5 MeV (million electron volts); thus the mass of the proton can be written as 1·0073 amu or 938·3 MeV/c^2.

Example: Find the binding energy per nucleon of the carbon nucleus, C^{12}.

Observed nuclear mass { Mass of carbon atom by definition = 12·0000 amu
Less mass of 6 electrons (ea. 0·00055 amu) → 11·9967 amu

Calculated nuclear mass { Mass of 6 protons = 6·0438 amu
Mass of 6 neutrons = 6·0522 amu

Sum = 12·0960 amu

Therefore the mass defect = (12·0960 − 11·9967) = 0·0993 amu = 93 MeV/c^2

∴ Binding energy per nucleon of C^{12} = $\frac{93}{12}$ = 7·75 MeV per nucleon

Figure 7.1 shows the binding energy per nucleon plotted against A for various nuclei. Apart from a few light elements such as helium (which we will refer to later) the trend is remarkably clear; a steep rise (region A) to a flat maximum (region B) slowly falling (region C) towards the heaviest elements. This diagram is of great importance to the problem of obtaining useful energy from nuclei. Clearly, to extract energy from a system of nucleons one must reduce their mass, that is, increase their total binding energy. This can evidently be done either by bringing together two light elements from region A to form a nucleus in region B (a process known as *fusion*) or by splitting a nucleus of region C into two fragments, each of them in region B (*fission*). It is believed that the energy of the sun comes from the first process; the precise details are obscure but something like the fusion of neutrons and protons to form carbon, as in the example above, is involved. Fission, specifically of uranium or plutonium, is the process that provides energy for atomic (strictly, 'nuclear') weapons and reactors. Uranium has several isotopes, U^{238} being the commonest. To induce fission the nucleus must somehow be excited, for instance by neutron bombardment. If fission occurs,

---

* The idea of binding energy is not restricted to nuclei. The total mass of an atom, for instance, is slightly less than the masses of the individual electrons plus nucleus. Similarly a liquid, such as water, possesses binding energy of the individual molecules; when this energy is supplied, for example by heating, the molecules will separate as steam. In the same way nuclear binding energies are a measure of the work needed to pull a nucleus to pieces, and this is perhaps the simplest way of thinking of it.

several neutrons are released in addition to the two nuclear fragments. The total energy released can be seen from Fig. 7.1 to be about 1 MeV per nucleon, that is about 200 MeV for a uranium fissio Each of the two *fission fragments* is therefore emitted with approxi mately 100 MeV of kinetic energy, and how this energy is finally di sipated is a problem we shall return to later. However, the energies of the neutrons are low, usually less than 2 MeV, and they are unab to excite a U^{238} nucleus sufficiently to cause fission. Thus natura uranium is non-fissile and quite innocuous in this respect.* Howeve the rare isotope U^{235} is much less stable, and when struck by even low energy neutron will undergo fission, releasing more neutrons in the process. Thus a *chain reaction* can be set up only in uranium which has been enriched in U^{235}. Given an arbitrary lump of such enriched uranium the chain reaction does not necessarily occur, be cause the excess neutrons produced in each fission might escape from the lump before they can initiate another fission. Only when the amount of uranium present is above a so-called *critical mass* can one be certain that at least one neutron, on average, is available to carry on the fission reaction. The whole art of nuclear reactor design is to ensure that 'criticality' is just reached but never exceeded; in designing nuclear weapons, of course, one attempts to bring together two (or more) *sub-critical* masses as quickly as poss ble, to exceed the critical mass.

Radioactivity

Fission is one process by which heavy elements can get rid of their excess energy. However, it is a process which requires some sort o initial excitation, and anyway is possible only for the heaviest elements. There are a large number of elements, most of them in regi C of the binding energy curve, which lose energy by spontaneous emission of various particles or radiations, and so move towards th more stable region *B*. This phenomenon is called radioactivity and it was discovered by a Frenchman, Becquerel, in 1896 only a few months after Roentgen's discovery of X-rays. The Curies, Pierre a Marie, were also involved at an early stage, and isolated the new element radium in July 1896.

Within a few years it became clear that three types of radiation were involved. From ignorance of their precise nature (as was the

---

* Uranium's greatest hazard lies in its being pyrophoric, i.e. spontaneously inflammable, when in powder form!

Nuclear Radiations

case with X-rays) they were named non-committally alpha (α), beta (β) and gamma (γ) rays. The names have stuck though simple experiments soon showed that their characters were quite distinct. An early experiment is shown schematically in Fig. 7.2. By using suitable detectors such as photographic film, which is blackened by these radiations, it was demonstrated that the α-rays were deflected towards the cathode, and hence were positively charged;

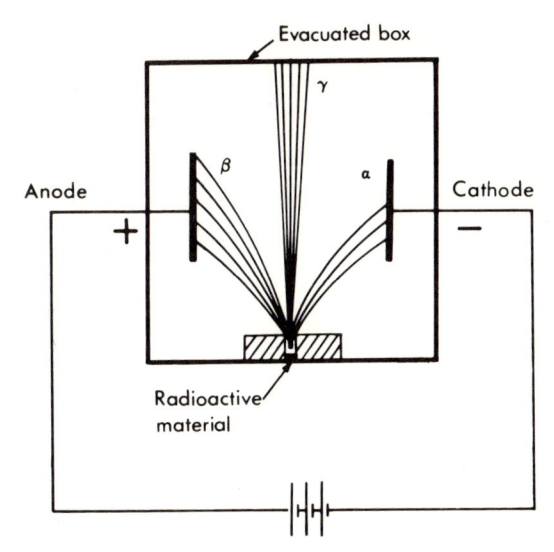

Fig. 7.2
The distinctive behaviour of α, β and γ rays in an electric field is illustrated by a schematic experiment.

the β-rays were deflected to the anode and were therefore negatively charged; whereas the γ-rays were quite undeflected, by either electric or magnetic fields. By such experiments it was proved that α- and β-rays were really streams of particles, and only γ-rays were true electromagnetic radiation; they are in fact similar to, though of much shorter wavelength than, X-rays.

Let us discuss α-, β- and γ-rays in turn, describing their nature and production process. Alpha-particles consist of a tightly-bound cluster of two protons and two neutrons, that is, they are nuclei of helium atoms. They are emitted entirely by elements heavier than lead, the removal of four mass units enabling these unstable nuclei to move towards the stable region of the binding energy curve. The reason that helium nuclei (and not, so far as we know, protons or deuterons) are emitted in this way is that helium itself is extremely stable; reference to Fig. 7.1 shows that its binding energy, at 7 MeV per nucleon, lies well above the general trend for low A. In fact, recent experiments give some indication that many nuclei actually contain 'clusters' of α-particles, more or less separated from one another. Alpha-decay is the emission of one such cluster.

The typical energies of α-particles are easy to estimate from the graph of Fig. 7.1. In the α-emitting region near $A = 200$, where the binding energy per nucleon is about 8 MeV, a change of 4 mass units is seen to change the binding energy by about $\frac{1}{20}$ MeV per nucleon. So a typical parent nucleus will have a total binding energy of $200 \times 8 = 1600$ MeV, and the decay products will have binding energies of $196 \times \left(8 + \frac{1}{20}\right) = 1578$ MeV (for the daughter nucleus) and $4 \times 7 = 28$ MeV (for the α-particle), a total of 1606 MeV. Thus 6 MeV is released as kinetic energy. In fact, specific decays range in energy release from 4 MeV to over 8 MeV. The precise amount is easily calculable if the initial and final masses are accurately known. For example, Ra^{226} (mass = 226·096 amu) decays to Rn^{222} (222·087 amu plus an α-particle (4·004 amu). The mass change is 0·005 amu which is equivalent to about 4·7 MeV. (Note that this energy is shared between the α-particle and the radon nucleus. In Chapter 1 this is taken as an example and the laws of conservation of momentum and energy are used to determine how much energy each particle takes.)

Turning now to β-rays, these are known to be identical with ordinary atomic electrons. Unlike alphas, they do not possess a fixed, calculable energy, but for a particular radioactive decay may have any energy from zero up to a well-defined maximum which is rarely greater than 2 MeV. Two problems that arise, then, are why beta ray do not have a discrete energy, and how they can be emitted at all by nuclei which consist only of neutrons and protons. The answer to the first question is clearly that β-decay, unlike α-decay, *cannot* be a so-called 'two-body' process, with only two objects produced. If it were then each would have a definite energy, as in α-decay. There must be a third body involved in β-decay. Now we can answer the second question, because it is known that neutrons are themselves unstable particles; left to itself a neutron will decay, to a proton, an electron and a third particle which is uncharged and so light that it is believed to be completely massless. This particle is called the *neutrino* (Italian for 'little neutral one') and its sole interest for us

Nuclear Radiations

is that it can carry away some of the energy that would otherwise be taken by the proton and electron. Symbolically we can write

$$n \to p + e^- + \bar{\nu} \qquad (7.1)$$

(where we have been pedantic and written a bar over the Greek letter ν, the symbol for the neutrino; this means it is actually an *antineutrino* that is involved). The neutrino has virtually no interaction with matter; it has been estimated that the number of neutrinos produced by nuclear reactions in the sun, and which pass through our bodies, is about 10^{11} per second, but this enormous flux has virtually no detectable effects.

From the known neutron, proton and electron masses we find that the maximum energy carried by the electron in reaction (7.1) is about $\frac{3}{4}$ MeV. In β-decay this reaction occurs inside the nucleus, and since nuclei with the same value of A rarely have binding energies more than $\frac{1}{2}$ MeV apart, we find a ready explanation for β-ray energies lying mostly within the range 0·5–1·5 MeV.

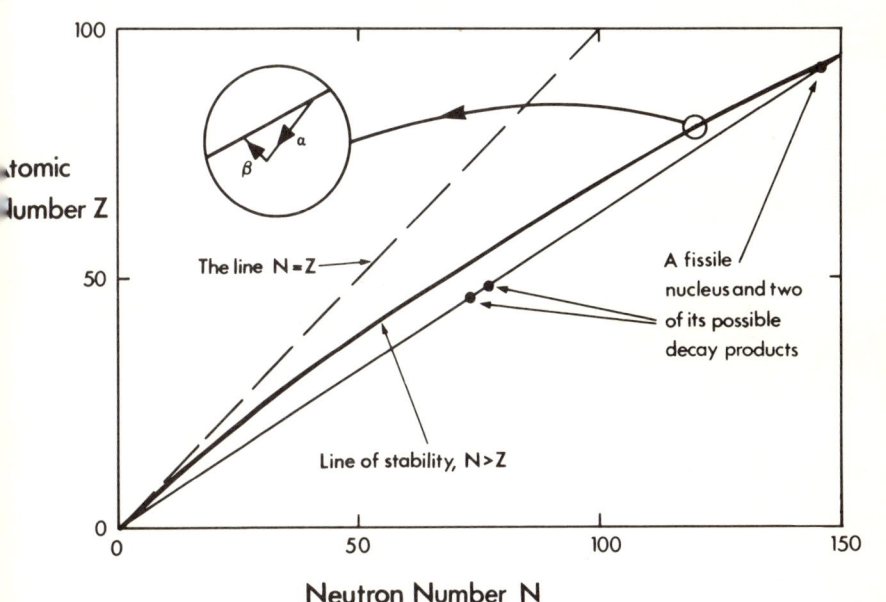

Fig. 7.3
Stable nuclei all lie close to the 'line of stability' shown in this graph of Z against N. The inset shows how successive α and β decays can result in stable nuclei, and on the main graph the instability of fission products against β decay is clearly shown.

To explain why β-decay occurs at all in nuclei we must qualify an earlier statement that nuclei contain roughly equal numbers of protons and neutrons. Heavy nuclei contain rather fewer protons than neutrons because the strong electrostatic repulsion between protons becomes more influential in large nuclei where the short-range nuclear binding force is less effective. In Fig. 7.3 we plot N against Z for naturally-occurring nuclei, and it is found that most nuclei lie near the 'line of stability' with $N > Z$. Now α-decay will generally leave the daughter nucleus below the line of stability (see inset, Fig. 7.3). To return to it, the nucleus must reduce N and increase Z, which is accomplished precisely in β-decay. Thus, one commonly finds β-decay following α-decay in heavy nuclei. Possibly the resultant nucleus is still unstable, and a whole sequence of α- and β-decays may occur before stability is reached. There are also a small number of isotopes of lighter elements, which lie off the line of stability and are therefore β-emitters. An example is the potassium isotope, K^{40}. Besides these there are many unstable isotopes which are produced in nuclear reactors (and explosions) as the result of the fission process. As is shown in Fig. 7.3, the splitting of a fissile nucleus into two roughly-equal fragments results in nuclei off the stability line which are very prone to β-decay. In uncontrolled fission, these nuclei are the source of radioactive *fall-out* from nuclear weapons; in controlled fission (reactors) these isotopes provide the source for most industrial and medical β-rays. A well-known example is Sr^{90} (β-ray energy 0·54 MeV).

Just as atoms possess discrete energy levels, so do nuclei. Either α- or β-emission may leave the daughter nucleus in an excited state from which it can reach the ground state by emitting a γ-ray. Typical energies are up to about 2 MeV for γ-rays following β-decay but only about $\frac{1}{2}$ MeV following α-decay. If the excited state is more than roughly 2 MeV above the ground state, it is likely to emit another β-ray rather than a γ-ray, and this sometimes occurs for very much lower energies also. As a rule, therefore, strong β-sources are generally γ-sources also.

Radioactivity is a *random* process. An early discovery was that the proportion of nuclei that decay in a given time is constant, being characteristic of the particular nucleus involved and unalterable by any chemical or physical processes. It is convenient to state this law mathematically; if, out of N nuclei, ΔN decay in a short time Δt, the law states that the fraction decaying, $\Delta N/N$, is proportional to Δt, or

Nuclear Radiations

$$\frac{\Delta N}{N} = -\lambda \Delta t \qquad (7.2)$$

the minus sign appearing because ΔN represents a decrease. Equation (7.2) is of a well-known type and can be solved to find how many out of the original number of nuclei, N_0, remain after time t. The answer is

$$N = N_0 \, e^{-\lambda t} \qquad (7.3)$$

where, as usual, $e = 2 \cdot 7183 \ldots$. The reciprocal of λ is often more useful than λ itself; we call it the *mean life*, τ, and it is obvious* that τ is the time after which a fraction $\frac{1}{e} = 0 \cdot 3679$ of the original nuclei are left. The *half-life*, $\tau_{\frac{1}{2}}$, is even more useful, being the time after which exactly half the original nuclei are left. $\tau_{\frac{1}{2}}$ is evidently shorter than τ, and it may be shown (try to do so) that $\tau_{\frac{1}{2}} = 0 \cdot 693 \, \tau$.

Natural radioactive elements exhibit a great variety of half-lives, including some extremely short ones, such as the 1·3 secs of At^{218} and the 160 μ secs of Po^{214}. The question arises, how such nuclei have survived in a universe that has existed for about 5×10^9 years? The answer is that practically all natural radioactive nuclei are members of one of three families, each descended from a very long-lived parent nucleus. These parents are U^{238} ($\tau_{\frac{1}{2}} = 4 \cdot 5 \times 10^9$ years), U^{235} ($\tau_{\frac{1}{2}} = 7 \times 10^8$ years) and Th^{232} ($\tau_{\frac{1}{2}} = 1 \cdot 39 \times 10^{10}$ years). Each family consists of a series of α- and β-emitters as described above, and the end product in each case is the stable element lead. A considerable amount of lead in the Earth's crust is of this *radiogenic* variety. One of the ways of determining the age of rocks is to measure the ratio between the amount of radiogenic lead and the amount of uranium (or thorium) they contain. A simple calculation then gives the time since the process started. A similar calculation is often possible with those few other natural radioactive nuclei which are not members of these three families — examples are K^{40} (decays to Ar^{40} with $\tau_{\frac{1}{2}} \simeq 1 \cdot 3 \times 10^9$ years) and Rb^{87} (decays to Sr^{87} with $\tau_{\frac{1}{2}} \simeq 5 \cdot 2 \times 10^{10}$ years).

Example: A pre-Cambrian rock contains 2000 ppm (parts per million) of Rb^{87} and 100 ppm of radiogenic Sr^{87}. What is its age?

---

* Substitute $\lambda = \frac{1}{\tau}$ in eqn (7.3) to yield $\frac{N}{N_0} = e^{-t/\tau}$ so at time $t = \tau$, $N/N_0 = e^{-1} = \frac{1}{e}$ = fraction of nuclei left.

From eqn (7.3) we find $\log_e \frac{N_0}{N} = \lambda t = t/\tau = \frac{0 \cdot 693 t}{\tau_{\frac{1}{2}}}$. Hence

$$t = \frac{\tau_{\frac{1}{2}}}{0 \cdot 693} \log_e \frac{N_0}{N} = \frac{4 \cdot 3 \times 10^{10}}{0 \cdot 693} \log_e \frac{2100}{2000} = 3 \cdot 6 \times 10^9 \text{ years.}$$

Besides its characteristic decay-products, it is frequently necessary to specify the strength, or *activity*, of a particular radioactive source, that is the number of disintegrations that occur per second. Of course, for a given number of atoms present, this can be calculated if the half-life is known, but it is frequently useful to combine these two quantities, and the unit that is now used is the curie (abbreviation Ci to distinguish it from C, coulomb). The curie is equal to a rate of $3 \cdot 70 \times 10^{10}$ decays per second, which is approximately the activity of a 1 gram sample of pure radium. The majority of sources used in laboratory experiments have low activities, often of only a few microcuries (1 μCi = 10^{-6} Ci) or even less. Sources used in therapy, however, are often extremely active, several thousand curies in some cases.

Example: What is the activity of 1 gram of U^{238}? 1 gr of U^{238} contains fewer atoms, in the ratio $\frac{226}{238}$, that 1 gram of Ra^{226}; moreover $\tau_{\frac{1}{2}}$ for U^{238} is greater than that of Ra^{226} in the ratio $\frac{4 \cdot 5 \times 10^9}{1600}$; therefore the activity of 1 g of U^{238} is 1 Ci $\times \frac{226}{238} \times \frac{1600}{4 \cdot 5 \times 10^9} \simeq \frac{1}{3} \times 10^{-6} = \frac{1}{3} \mu$Ci.

This completes our summary of radioactive decays. There is one further source of radiation in nature, and that is the cosmic radiation. This consists of particles, usually protons or photons, which are produced in the galaxy at immense distances from the Earth and are then accelerated to very high energies, by some mechanism that is not fully understood. On their striking the atmosphere various nuclear reactions occur, and only the resulting mixture of photons, electrons and unstable *mesons* actually reaches the ground. Life on Earth has presumably always been exposed to this cosmic radiation. Although its intensity varies slightly from place to place and year to year, a useful rule of thumb is that one particle passes through an area of 1 cm² every minute.

One side effect of cosmic radiation is the production of an unstable carbon isotope, C^{14}, by collisions with the oxygen and nitrogen of the atmosphere. C^{14} has $\tau_{\frac{1}{2}} \simeq 5600$ years and is incorporated, along with

Nuclear Radiations

normal C^{12}, into living plants. After their death the C^{14} content diminishes as it decays away, and measurements of this content are often used to assign a date to trees, wooden artefacts, paper — in fact anything containing vegetable matter.

The interaction of radiation with matter

All types of radiation lose energy when they pass through matter. The transfer of energy can take various forms; typically, however, direct *ionisation* of the material is the most important, though there are several other processes which can also result in energy loss. (We shall deal with some of these in this section, but will remark now that in most cases the end-result is to cause ionisation somewhere in the material.) The ionised atoms and molecules eventually return to their ground state, their energy either appearing as random molecular motion (heat) or emission of visible radiation (light), or being absorbed in endothermic chemical reactions. We shall discuss these effects in greater detail in later sections. Here we shall describe how ionisation and other forms of energy loss occur, and outline the differences between the main types of radiation. As we shall see, these differences are very important when considering the biological effects of radiation.

We begin with electromagnetic radiation such as X-rays and γ-rays. The characteristic feature here is that when such radiation passes through a slab of material the energy of each of the emerging photons (quanta of radiation) is unchanged but the total number of photons is reduced, that is the intensity of the radiation has been *attenuated*. The physical explanation is that virtually all the interactions that the photon makes change, not only its energy, but its direction as well, often very considerably so that the photon is completely removed from the beam. This will become apparent when we discuss each type of interaction (there are three main ones) in turn. Without going into details, though, we can state the attenuation law as follows. The probability that an interaction will occur is proportional to the distance travelled in the material, and if we let ΔI be the (small) change in intensity, I, on travelling a distance Δx, this can be written

$$\frac{\Delta I}{I} = -\mu \Delta x \qquad (7.4)$$

where μ is some constant known as the *attenuation coefficient*.

Note that eqn (7.4) is of exactly the same form as eqn (7.2) and therefore leads to the same exponential law (compare eqn 7.3).

$$I = I_0 e^{-\mu x} \qquad (7.5)$$

The quantity μ/ρ, where ρ is the density, is called the *mass attenuation coefficient*, and is more informative than μ as it is found in practice that μ/ρ has very similar values whatever the material used. I_0 is the initial intensity, where $x = 0$.

The three main types of interaction are the *photoelectric effect*, *Compton scattering*, and *pair production*. Their relative importance depends on the energy of the γ-ray (see Table 7.1, p. 190). At low energies ($E < 0.5$ MeV) the photoelectric effect dominates. In this the whole of the γ-ray energy goes to ionising an atom, any excess being carried away as kinetic energy by the electron. Thus the γ-ray is completely absorbed in this process, though the ejected electron will move through the material, causing further ionization until its energy falls below the ionisation energy of the material. The attenuation coefficient for the photoelectric effect falls rapidly with increasing energy; it is fairly evident that for a given energy E, the effect will be more significant for materials with many bound electrons, that is with high Z. It can in fact be shown that $\mu \propto Z^5$ for the photoelectric effect.

For higher energies the process of Compton scattering (named after the American physicist who studied it in the 1920s) becomes more important. In this process the photon is not absorbed but scatters elastically off the electrons in the material; the photon recoils with reduced energy and the struck electrons can cause further ionization until their energy becomes too low. Again the attenuation coefficient increases as the number of electrons goes up, and it is easy to show that $\mu \propto Z$. As regards the energy-dependence of μ, the scattering process we have described is most efficient when the γ-ray energy is equivalent (using $E = mc^2$) to the mass of the electron, that is 0.51 MeV. Thus μ reaches a maximum around 0.5 MeV, and decreases rather slowly at higher energies.

The photoelectric effect and Compton scattering both involve interaction with electrons.* The third process we shall mention

---

* One reason that nuclear collisions are unimportant for γ-rays is that the latter's wavelength λ, as given by $E = h\nu = \frac{hc}{\lambda}$, is usually much greater than nuclear dimensions. Thus a 1 MeV γ-ray has $\lambda = hc/E = \frac{(1 \cdot 24 \times 10^{-10} \text{ MeV cm})}{1 \text{ MeV}} \approx 10^{-10}$ cm. Since most nuclei have diameters $\approx 10^{-12}$ cm, λ is about 100 diameters. Nuclei are as inefficient at scattering γ-rays as minute dust particles are at scattering light.

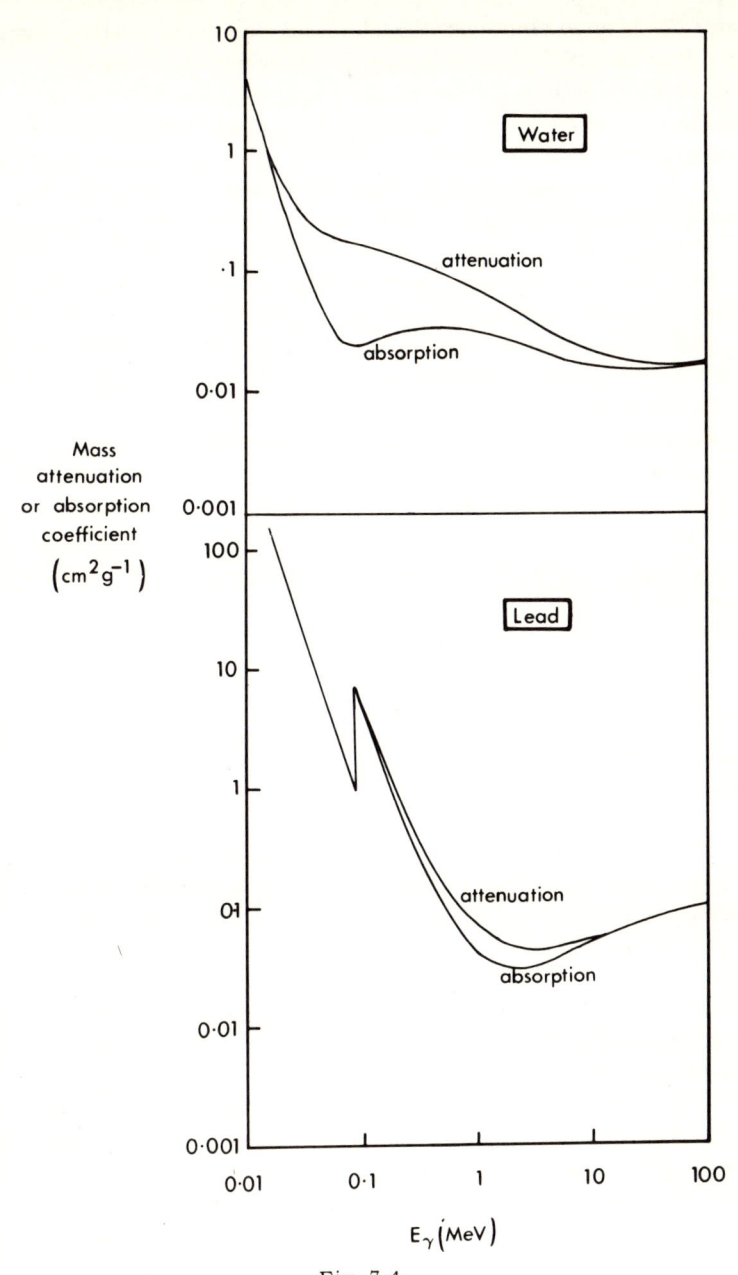

Fig. 7.4
Absorption coefficient as a function of energy for γ-rays in lead and in water (which approximates to biological tissue). The scales are logarithmic.

does not involve electrons (though it can only occur in matter, not in a vacuum). If E is greater than $2m_e c^2$, where m_e is the electron mass, the γ-ray can convert all its energy into the mass of two electrons (one an ordinary negative electron and the other a positive electron or *positron*, so that electric charge is conserved). This process of pair production therefore starts at $E = 2m_e c^2$ and becomes at higher energies the dominant mechanism by which γ-rays lose energy. The pair of electrons carry on ionising in the usual manner, though the positron will quickly annihilate with a negative electron to produce two more γ-rays.

In Fig. 7.4 we show how the mass attenuation coefficient varies with γ-ray energy for two typical absorbers, water (approximately the same as biological tissue) and lead.

Example: How much lead is required to reduce the intensity of γ-rays of 0·5 MeV by a factor of 1000? How much water is required for the same effect? By what factor are such γ-rays attenuated in passing through an average human body?

From the graphs, $\frac{\mu}{\rho} = 0\cdot 15 \text{ cm}^2 \text{ g}^{-1}$ for lead, for which $\rho = 11\cdot 3$. Substituting in eqn (7.5) we find $x_{Pb} = \log_e (I_0/I) / \left(\frac{\mu}{\rho} \times \rho\right) = \frac{6\cdot 9078}{0\cdot 15 \times 11\cdot 35} \simeq 4\cdot 0 \text{ cm}$. For water, $\frac{\mu}{\rho} = 0\cdot 07 \text{ cm}^2 \text{ g}^{-1}$ so similarly $x_{H_2O} = \frac{6\cdot 9078}{0\cdot 07 \times 1} \simeq 100 \text{ cm}$. If the human body has $x \sim 30 \text{ cm}$, the attenuation factor is $(I_0/I) = \exp(\mu x) = \exp(0\cdot 07 \times 1 \times 30) = \exp(2\cdot 1) = 8\cdot 17$.

In contrast with the complex behaviour of photons, charged particles lose energy almost entirely by collision with atomic electrons and consequent ionisation of the atom. High-energy β-rays may also collide with nuclei and emit γ-rays; this *bremsstrahlung* process is analogous to X-ray production. First, we shall discuss the important heavy particles such as α-particles, protons and fission fragments. The passage of such particles through matter is analogous to that of a heavy cannon ball through a flock of small birds; the particle slowly loses energy through a large number of collisions, but is not deviated appreciably from its initial direction. The energy lost in each collision is approximately the energy needed to ionise the atoms, and for most absorbing materials this is about 30–40 eV. Thus a 3 MeV proton undergoes about $\frac{3 \times 10^6}{30} = 10^5$ collisions before it is brought to rest. Because of this

Nuclear Radiations

very large number of collisions we can treat the whole problem statistically (as we did when discussing diffusion in Chapter 2) and can calculate both the energy lost in travelling unit distance in the material (the *stopping power*) and the total distance travelled (the *range*) for particles of various energies. The greater the electric charge of the particle, the greater its ionization per unit path length and so the shorter its range. Thus, for example, protons of 10 MeV have a range in air of about 112 cm, whereas α-particles of the same energy travel only 10·5 cm. (As a useful rule-of-thumb, an α-particle of E MeV will travel E cm in air.) Materials denser than air will stop these particles in correspondingly shorter distances, and it is true to say that no naturally-occurring α-particles will penetrate an ordinary postcard. Conversely the stopping power for these particles is very high, which is another way of saying that the density of ionization along their path is high. Moreover, the stopping power increases as the energy decreases, so that the density of ionization is greatest at the end of a heavy particle's range. In the frontispiece we show a photograph of the tracks of various particles in photographic film, the degree of blackening being proportional to the amount of ionisation. Note the dense, short α-particle tracks, and the even longer track due to a proton emitted in a nuclear reaction. For comparison, note how lightly and irregularly ionized are the tracks of the β-rays. (The γ-rays, of course, do not ionize; we see only the tracks due to recoil electrons and electron-positron pairs.)

The energy loss of β-rays is more akin to that of γ-rays than to heavy particles, in that individual collisions with atomic electrons deflect the β-ray very considerably. The energy lost in each collision is highly variable and depends on the angle through which the β-ray is scattered. For these reasons the actual path of an electron through matter is highly convoluted (akin to the 'drunken walk' problem of Chapter 2) and, although the total range measured along this path is constant for a particular energy, the actual distance of penetration into the material can vary considerably from one β-ray to another. Consequently electrons obey an exponential absorption law like eqn (7.5), the *mean range* being the distance within which one-half of the electrons are absorbed. Electron ranges are very much greater, for a given energy, than those of heavy particles. For example, an 0·1 MeV β-ray has a mean range of about 10 cm in air, roughly 100 times that of an 0·1 MeV α-particle. The reason is simple; the stopping power, and hence the range, is

TABLE 7.1
Processes by which particles lose energy in matter

Particle	Process	Energy region	Characteristic features
Photons (X-rays & γ-rays)	Photoelectric effect	$\lesssim 0.5$ MeV	Exponential attenuation ($I = I_0 e^{-\mu x}$)
	Compton scattering	0–10 MeV, roughly (max. around 0.5 MeV)	Ionisation roughly constant along path of photon beam
	Pair production	> 1.0 MeV	
β-rays (electrons)	Scattering by atomic electrons	All energies	Range *along* path is fixed but path is convoluted leading to effectively exponential attenuation and roughly constant ionisation
	Bremsstrahlung	High energies (mainly above 1 MeV)	
Protons α-particles, fission fragments	Direct ionisation of atoms	Ionisation increases as energy gets less	Fixed range, greatest for particles of lowest mass and charge. Ionisation greatest near end of range
Neutrons	Nuclear interaction e.g. $n + p \to d + \gamma$ $n + B^{10} \to Li^7 + He^4$	'Thermal' neutrons ($\sim \frac{1}{40}$ eV)	Much longer range than charged particles
	Scattering off nuclei leading to recoil fragments	All energies	Random but intense patches of ionisation along path

Nuclear Radiations

determined by the particle's velocity, since the greater the time spent in travelling unit distance, the more ionizing events are likely to occur. Now particles of the same (low) kinetic energy have equal values of $\frac{1}{2}mv^2$, so their velocities are in the inverse ratio of the square roots of their masses. For alpha-particles and electrons this ratio is $1 : \sqrt{7200} = 1 : 85$, which is fairly close to the observed ratio of ranges. (This way of accounting for the range discrepancy is correct at all energies, but at high energies the fact that the electron is moving with a velocity close to that of light must be taken into account; it results in the discrepancy being somewhat less at high energies.) The attenuation of β-rays is summarized in Table 7.1.

The remaining biologically important particles are neutrons. Being uncharged they cause no direct ionization but lose energy only when they interact with nuclei in the absorber. Neutrons are thus attenuated exponentially, and since nuclei present a much smaller target than do the atomic electrons, neutrons typically have long mean ranges, that is small absorption coefficients. Many feet of a cheap, dense material such as concrete are needed to provide efficient shielding against neutrons. The nature of the nuclear reactions depends on the neutron energy. At low energies (so-called *thermal neutrons* travelling with kinetic energy $\frac{1}{2}mv^2 \approx kT$ have an energy $\approx \frac{1}{40}$ eV) neutrons may collide with free protons (hydrogen nuclei) and form the isotope deuterium, releasing the deuteron binding energy of 2·2 MeV as a γ-ray. The γ-rays are then attenuated as usual. Faster neutrons, colliding with nuclei such as oxygen, can ionise them completely, the recoil nuclei behaving as do fission fragments and causing dense ionisation.

Detection of radiation

Nearly all methods of detecting and measuring radiation make use of its ionising action in passing through matter. Thus it is easy to detect charged particles which ionise heavily, relatively easy to detect electromagnetic radiation which, though not itself ionising causes copious secondary ionization, but rather difficult to detect uncharged particles such as neutrons, whose ionising effects are sparse and unevenly distributed.* In this section we shall describe the main de-

---

* Neutrinos are almost impossible to detect by any conventional means, but since this is because of their negligible interaction with matter in any case, the point is of no consequence to the health physicist.

tection techniques for charged particles, γ-rays and neutrons. It is convenient to divide these techniques into three, utilising respectively ionisation in a gas, ionisation in a solid or liquid, and speci nuclear interactions (for the detection of neutrons).

Detection by measuring the ionization produced in a gas is histo cally very significant, for it was by these means that both X-rays a cosmic rays were first detected. Physicists often make use of an i strument called an electroscope (Fig. 7.5a) which consists of two thin gold foils attached to a metal rod and placed in an earthed and insulated box. When the rod is electrically charged to a high poten (voltage) the foils both become positively (or negatively) charged a diverge due to their mutual repulsion, the angle θ that they make w each other being a measure of their potential. Now it was found tha these electroscopes, no matter how carefully constructed and in sp of stringent precautions to prevent electrical leakage through the s ports, always discharged themselves slowly, indicating some elect leakage through the air in their boxes. This was an indication that

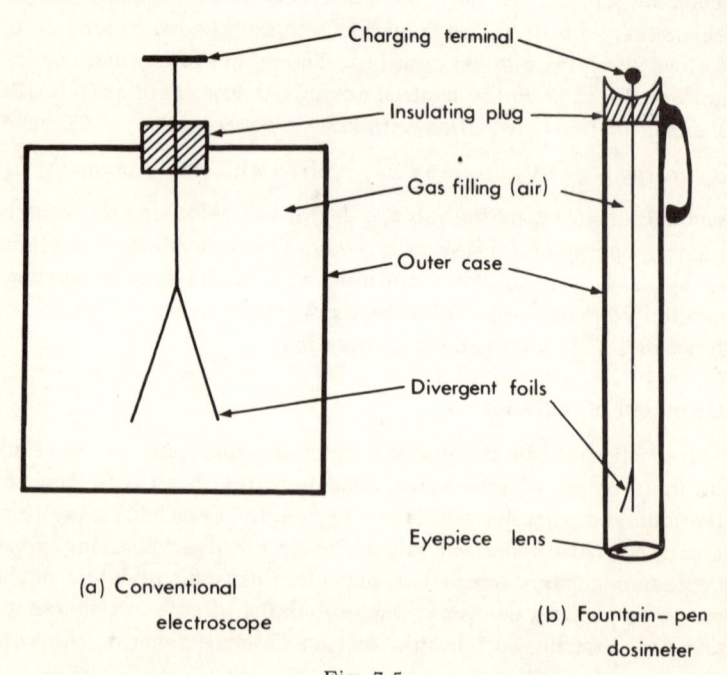

Fig. 7.5
Two versions of the gas counter. (a) Conventional gold leaf electroscope. (b) Modern 'fountain-pen' dosimeter.

some pervasive radiation was present, ionising the air in the electroscope and thus allowing the ions to drift across between the box and the foils, conducting the charge away from the latter. (That cosmic radiation was indeed extra-terrestrial was indicated by the fact of this leakage current increasing when electroscopes were carried to mountain tops and thus raised above much of the attenuating layers of the atmosphere.)

In only a slightly modified form the electroscope is still used to monitor ionising radiation; the device is reduced in size and, shaped like a fountain-pen, can be carried in the pocket (Fig. 7.5b). The divergent foils can be viewed end-on through a lens at one end of the 'pen', and at the other end is the terminal for recharging the instrument. A graduated scale is incorporated so that this *fountain-pen dosimeter* can be calibrated against sources of known strength, to give absolute values of amount of ionization received. The dosimeter is sensitive to β- and γ-rays but not α-particles (which cannot penetrate the walls) nor neutrons.

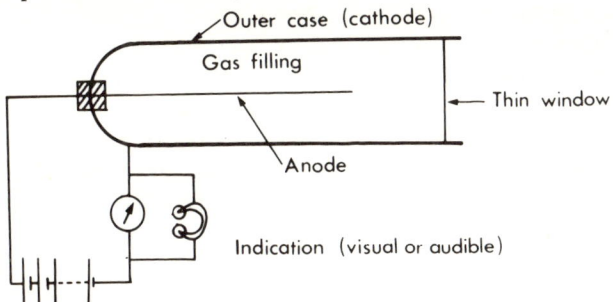

Geiger-Müller counter (schematic)
Fig. 7.6
Principle parts of a Geiger–Müller counter.

Such an instrument is used for recording the effects of radiation integrated over a period of time. Devices which respond to the passage of individual particles are also required, and we shall describe one such instrument, the Geiger–Müller counter, which relies on gas ionisation for its operation. In its simplest form this consists (Fig. 7.6) of a gas-filled envelope, its metal case forming the cathode, with a thin axial wire charged to a high potential forming the anode. The gas, usually a noble gas such as helium, neon or argon, is at a pressure of about $\frac{1}{100}$ th of an atmosphere and stressed by the electric field (typically 500–1000 volts per cm) until it is on the point of breaking down and causing a spark. When a charged particle or γ-ray enters the counter the electrons produced are accelerated towards the electrodes and quickly reach an energy sufficient to cause further ionization. The process is repeated with the secondary electrons and an *avalanche discharge* develops very rapidly (in less than one millionth of a second). The effect is that the gas, which was previously a very good insulator, has suddenly become conducting because of the very large number of ions and electrons it now contains, and consequently there is a very large pulse of current in the circuit between anode and cathode. This pulse can be made to operate earphones or a loudspeaker to produce an audible click, or it may be amplified electronically and give a visual indication, using small neon lights, say, of the total number of pulses that have been recorded. (To avoid confusion with the detecting instrument itself, which is always called a *counter*, this operation of registering the number of counts is called *scaling*, and the instrument that performs it is a *scaler*.)

The current pulse which has developed must be stopped somehow and this is usually done by adding to the gas filling a small quantity of a substance (alcohol, for example) which is easily dissociated by the energy contained in the avalanche discharge. The dissociation absorbs the energy and stops the discharge, preparing the counter for the next particle. The irreversible dissociation of this *quenching agent* sets a limit to the useful life of such a Geiger counter. (More sophisticated counters have a quenching action built-in to their external circuitry, and thus have a longer life.)

The Geiger–Müller counter has the following advantages over the fountain-pen dosimeter. Individual particles can be counted, and the response is so rapid that very high rates can be recorded, limited only by the recovery time (quenching time) of the counter. Thus it is used when an accurate and immediate indication of radiation levels is required. By using thin windows counters can be made which are sensitive to α-particles; also by careful design the response to γ-rays of low ionisation can be very much reduced. Thus it is possible to use these counters to distinguish between the various types of radiation.

Turning now to ionization in solids, the Geiger principle of detection by causing an avalanche discharge has only very recently been applied successfully (in so-called '*solid-state*' *counters*) and we shall not discuss it here. Most detection devices rely on either the chemical reactivity of the ionisation products, as in the photographic process, or on the emission of light as excited molecules decay to their ground states (the *scintillation* process). The first, like the fountain-pen dosimeter, usually records radiation over relatively long periods of time, and the second can be used to count and identify individual particles. Both techniques have been used since the earliest days of radioactivity, the fogging of photographic film indeed contributing to its discovery. The earliest scientific study of α-particles was made by Rutherford, who detected them by looking at the bright flashes of light they produce on striking a screen of a suitable phosphor (zinc sulphide was used in these early experiments).

Photographic emulsions are now available which are sensitive to all types of radiation (except α-particles which of course cannot penetrate the necessary light-tight wrapping). Even neutrons can be recorded, albeit with low efficiency, in certain emulsions. The *film badge*, a small piece of film clipped in a case and pinned to the clothing, is a familiar sight on all those who work with radiation sources, including X-ray machines. The film is developed at regular intervals and the degree of blackening is a record of the amount of radiation received in that time.

The use of light-sensitive counters has developed rapidly since the days when Rutherford used a microscope to count visually the number of scintillation flashes on the screen of his so-called 'spinthariscope'. For instance, these flashes were produced only on the surface, since, even if the particles are penetrating, the screen itself is opaque to light. Now, one uses transparent materials (particularly plastics and organic liquids) so that light can be collected from the whole path of the ionizing particles. This light is collected and amplified by light-sensitive vacuum tubes (like radio valves) called *photomultipliers*, and the count is recorded on a scaler. The response of these *scintillation counters* is even more rapid than that of Geiger–Müller counters, and they can therefore count accurately very high levels of radiation. Their great sensitivity makes very large counters possible; however, their main advantage lies in the fact that the amount of light given out is roughly proportional to the particle's energy, whereas Geiger–Müller counters give out a standard pulse for particles of any incident energy. Moreover, since scintillating materials are solid (or liquid) there is sufficient material to stop particles of very great energy indeed. In particular, even γ-rays of very small attenuation coefficient can be

TABLE 7.2
Devices for detecting radiation

Device	Sensitive to	Main uses and features	Type
Fountain-pen dosimeter	β and γ rays	Short-period ($\sim$ hours) personnel monitoring. Rechargeable after use. Gives instantaneous reading	Recording long-term levels of radiation; not usually used to distinguish individual particles
Film badges	β and γ rays; neutrons (inefficient)	Long-period ($\sim$ days) personnel monitoring. Non-rechargeable. Can only be read after use	
Geiger-counter	α, β, γ rays $\big\{$ can be selected to suit application	Fast, robust; visual or audible indication	For recording individual particles and accurate counts of high levels of radiation
Scintillation-counter	α, β, γ rays; neutrons $\big\{$ can be selected to suit application	Fast, versatile. Visual indication only (usually). Requires care in use. Bulky accessories	
BF$_3$ neutron counter	Thermal neutrons (also fast neutrons when used with a moderator)	Bulky, otherwise as for Geiger counter	

counted efficiently. Two disadvantages of these counters are that the photomultiplier tubes are fragile and require very high operating voltages; the simpler and more robust Geiger counters are therefore often preferred except for accurate laboratory work.

Finally, we mention methods of detecting neutrons. Most of these utilise one or other of the nuclear reactions which can be induced by neutrons; for instance, there is a very large probability for slow (thermal) neutrons to be captured by the boron isotope B^{10}, which immediately splits up (by α-decay) into Li^7 plus an α-particle, He^4. These heavily-ionising fragments are easily detected in a conventional Geiger counter. Usually the boron is in the form of its fluoride, BF_3, which is a gas and is used to fill the Geiger counter. If neutrons of higher energies need to be counted they must be slowed up before a BF_3 counter is efficient, and the counter is then encased in a thick paraffin-wax *moderator*, which slows the neutrons by collision with the hydrogen nuclei. Such neutron counters have a characteristic massive globular shape. Other counters relying on different neutron reactions have also been developed.

These various methods of particle detection are summarized in Table 7.2.

Biological effects of radiation

It is generally considered that the effects of radiation are harmful to living organisms. We shall discuss in a later section various instances where radiation can be beneficial, to mankind at least; here let us just say that even the generalisation is not entirely true. If the radiation has insufficient energy to cause ionization, as is the case with radio waves, there are no known effects, either good or bad, on living tissues. In the case of visible light, there may be sufficient energy in each quantum to cause molecular dissociation, but the body has learnt to adapt to this and turn it to good use, in the eye. Neutrinos are different again; though present in large numbers they do not interact at all.

However, the types of radiation we have been dealing with in this chapter all lose energy by ionization (or nuclear reactions in the case of neutrons) and can therefore produce reactive ions such as OH, O_2H, etc. Because living cells require such a delicate chemical balance to function properly, a small disturbance to this balance can have very large-scale effects, and if cell catalysts such as enzymes are involved these effects can multiply rapidly, and are virtually always deleterious. An example, so widespread that it is often not considered as a radiation effect, is *sunburn*. An excess of ultra-violet light causes dissociation of water molecules beneath the skin, and the resultant OH radical initiates a chain of reactions leading to all the common symptoms of sunburn.* (Note that this property of ultra-violet light is what usually *protects* us from its effects; the rays are strongly absorbed by water vapour in the atmosphere, and it is only on clear days with no haze that this 'filtering' effect is diminished.)

Besides this direct damage to tissues, radiation can produce long-term effects through genetic alterations (mutations) to the DNA and RNA of the germ-cells. These two types of deleterious effects are both termed *radiation damage*. Without going into details one may make the following general observations about radiation damage to organisms:[†]

(a) There is a wide difference in the radiosensitivity of different tissues.

(b) Within a particular tissue or cell, different regions again vary widely in radiosensitivity; for example, the cell nucleus and membrane are considered more sensitive than the cytoplasm.

(c) As a rule, the most radiosensitive cells are those undergoing rapid reproduction, that have a large mitosis (fission) time, and that are least functionally differentiated. Thus a child (and even more so a developing foetus) is more sensitive than an adult.

(d) The body can repair all radiation damage (except genetic mutations in germ cells) given time to do so. Thus equal doses will be less damaging if received over longer periods of time (just as is the case with sunburn!).

As would be suggested by (c), the most radiosensitive tissues are the gonads, lymphatic organs, bone marrow, skin epithelium and mucous membrane of the gut, while the least sensitive are the brain, central nervous system, and striated muscles.

Radiation dosimetry

In order to make meaningful assessments of the damaging effects of various types and intensities of radiation, it is necessary to quantify

---

[*] More complex effects are involved when lower organisms are killed by ultra-violet light. See the reference cited in the bibliography.

[†] The majority of studies of radiation damage have been carried out on lower animals. Direct knowledge of effects on humans has been gained either from very low-level, or localized, experiments, or from accidental exposures. Perhaps as a consequence of one's unwillingness to experiment on humans, the subject of radiation damage, like that of experimental psychology, is imperfectly understood, and only general statements can be made with confidence.

Nuclear Radiations

what we mean by *dose*, and the relation of this concept to that of damage. Unfortunately a rather confusing variety of units has been used at one time or another; here we restrict ourselves to those most commonly met. The original impetus to this subject was given by X-ray workers who measured intensity levels using air ionization chambers, and the original unit of radiation dose, the *roentgen*, was defined as the amount of X-rays (or γ-rays) required to produce 2.08×10^9 ion-pairs per cc of dry air at STP. This cumbersome definition was related to the current that could be drawn from ionization chambers of a particular volume. It is more convenient to redefine the roentgen in terms of the amount of energy deposited per gram of material, and this can be done using our knowledge that the energy necessary to produce one ion-pair in air is about 32.5 eV. Thus 1 roentgen = $(32.5 \text{ eV/ion pair}) \times (2.08 \times 10^9 \text{ ion pairs/cc}) \times (1.6 \times 10^{-12} \text{ ergs/eV}) \div (0.0013 \text{ g/cc})$

$\qquad$ = 83·3 ergs/g of air

$\qquad \simeq$ 93 ergs/g of water (or biological tissue)*

The difference between air and water is due to the difference in the average energy to produce one ion-pair in these two materials. A more convenient unit which is slightly greater than and has almost entirely replaced the roentgen is the *rad* (radiation absorbed dose), defined as the amount of X- or γ-rays required to deposit 100 ergs/g in the particular absorber considered (e.g. biological tissue).

Knowing both the activity of a particular source, and the energy of γ-rays emitted, one can evaluate the *dose-rate*, that is the dose received by an absorber in unit time.

Example: Na²² emits γ-rays of 1.277 MeV. If one accidentally swallowed the contents of a capsule containing $1\,\mu$Ci of Na²² in saline solution, what would be the dose-rate, averaged over the whole body? Assume the absorption coefficient is such that one-half of the total energy of the γ-rays is deposited in the body, which has a mass of 65 kg. The total energy absorbed =
$$\frac{(1 \cdot 277 \times 10^6)}{2} \times (3 \cdot 70 \times 10^4) \text{ eV per second.}$$ Hence the dose-rate =
$$\frac{\frac{(1 \cdot 277 \times 10^6) \times (3 \cdot 70 \times 10^4) \times (1 \cdot 6 \times 10^{-12} \text{ ergs/eV})}{2}}{6 \cdot 5 \times 10^4 \text{ grams}}$$
$= 0.58 \times 10^{-6}$ ergs g^{-1} sec^{-1} = 0.58×10^{-8} rads sec^{-1}.

---

* On this page we use, for historical reasons, an obsolete unit of energy, the erg, equal to 10^{-7} J.

Note that this definition of *dose* applies only to X- and γ-rays. Since the important factor causing radiation damage is the rate of ionization along the track of a particle (sometimes called the *linea energy transfer* or LET, and measured in eV per cm or ergs per cm) the actual *damage* caused is higher for, say, an α-particle than for a γ-ray of the same energy. To take this into account a quantity called the *relative biological effectiveness*, or RBE, is defined. The RBE is a measure of the relative damage caused by a particular type of radiation, compared with that caused by γ-rays. (Although the RBE does go up as the linear energy transfer increases, the two are not directly proportional; the LET of fission fragments, for example can attain enormous values, but this intense ionisation occurs in such a small region that the overall biological effect is not as large as might be supposed.) The RBE of γ-rays is taken to be unity. The *damage* or effective dose is now defined as the dose (in rads) multiplied by the RBE; the units of damage are called *rem* (roentgen equivalent in man), i.e.

effective dose in rem = dose in rads × RBE

Table 7.3 lists typical RBE values for various types of radiation.

TABLE 7.3
The RBE of various types of radiation

	Typical spread	Average value
X-, γ-, β-rays	1	1 (by definition)
Thermal neutrons	1–5	3
Fast neutrons and protons ($\sim$10 MeV)	5–10	8
Deuterons	10–15	> 10
Alphas (4 MeV–8 MeV)	10–20	15
Heavy ions, fission fragments	20–200	> 20

Having defined the basic units of dosimetry, let us consider the dose necessary to cause various degrees of radiation damage in man and compare these doses with those actually received by the population at large. We shall take as a particular example the effects of total body exposure to a massive and sudden dose of γ-rays and neutrons, such as might be experienced in a nuclear explosion or in reactor accidents. As has been mentioned, the precise effects of radiation vary according to the organs involved and the type of radiation; the figures given here, though possibly typical of other situations, are specific only to such acute whole-body exposure. Table

Nuclear Radiations

7.4 indicates the clinical effects of increasing doses of radiation. Though the figures quoted are approximate, it is known that organisms of a particular species are remarkably uniform in their sensitivity to radiation, and it is believed that there is no more than a factor of two to three difference between individuals in their radiation tolerance. Because of this uniformity, a fairly precise measure of the radiation tolerance of an organism is the dose which causes death in half the exposed population. In man, this dose is about 400 rem. For most bacteria it is about 50 000 rem, and the dose required to inactivate viruses is of the order of 10^6 rem.

It is not known whether all radiation damage is, like genetic damage, cumulative in its effects, or whether only some is. There is a

TABLE 7.4

Clinical effects of a sudden whole-body dose of γ-rays and neutrons

Dose in rem	Effect
25	Lowest dose clinically detectable by lymphocyte count in blood
100	Damage to gut and bone marrow, leading to nausea and fatigue in about one-half of the exposed population
200	Certain nausea; loss of body-hair; possible death
400	Death of half the exposed population
600	Virtually certain death
> 2000	Some irrationality before death, possibly due to brain damage

school of thought that there exists a *threshold dose* below which the effects of radiation are non-injurious. According to certain workers, there is some slight evidence that very small doses may actually be necessary for life, perhaps by stimulating the organism in some way; for example, experimenters studying frogs' hearts kept alive *in vitro* found that it was essential to add small amounts of radioactive potassium to the nutrient solution to keep the hearts beating. However, the official, and safest, philosophy is that all radiation is injurious. For those whose occupation requires that they work in an environment containing radiation, upper limits for exposure have been set, termed *maximum permissible doses*. As an example, the maximum permitted whole-body exposure is a dose of 5 rem in any one year. Because of the body's recuperative powers, larger proportionate doses are allowed for short periods, as long as the annual dose is not exceeded. For instance, the maximum extremity (hands and feet) dose is 50 rem per year, with a short period limit of 25 rem per quarter-year. For comparison, Table 7.5 lists the annual dose received from various sources in the environment; the total environmental dose, including diagnostic X-rays, is seen to be ~ 500 mrem/year, about one-tenth of the limit set for radiation workers. On the other hand the occupational exposure in most nuclear laboratories is kept to a very small fraction of the permissible limit.

The effects of *chronic* (as opposed to *acute*) exposure to high levels of radiation of various types have been studied in some detail. For instance, early X-ray workers often suffered from skin cancer

TABLE 7.5

Radiation doses received by the whole population

	Sources	Approximate annual dose	Comments
Natural sources	Cosmic radiation	35 mrem/yr	Varies with latitude
	Environmental activity (U, Th, K^{40} in soil and rocks)	70 mrem/yr	Varies with location; up to 1000 mrem/yr in parts of India and Brazil
	K^{40} in body	20 mrem/yr	
	Ra^{226} in body	5 mrem/yr	
	Rn^{222} in air	~ 50 mrem/yr	Highly variable
Artificial (man-made) sources	Diagnostic X-rays	200 mrem/yr	Western hemisphere average
	Radioactive fall out	100 mrem/yr	Decreasing at present (1970)
	Luminous dials (watches etc)	2 mrem/yr	
	Fluorescope examination e.g. shoe fitting	50 000 mrem	Now discontinued — very dangerous
Total (excluding fluoroscope dosage)		~ 482 mrem/yr	Probable range for majority of population is 200–600 mrem/yr

caused by the heavy doses they unwittingly received. Another notorious example is that of the women dial-painters in the USA who applied luminous paint containing radium to instrument dials during World War I. They were in the habit of 'pointing' the brushes with their tongues, and the majority have now succumbed to cancer of

Nuclear Radiations

the stomach. This case emphasises the great danger of ingested α-sources, which can cause immense local damage due to their high RBE. β-emitters are generally less dangerous; there is no record of anyone being killed by surface β-ray exposure, though skin burns and hair-loss can be caused. One major problem is the high β-activity associated with many fission products; when these occur in radioactive fall-out they can be ingested. Sr^{90}, for instance, has a long half-life and can replace calcium in bones, with insidious effects. Fast neutrons are another dangerous radiation; with their high RBE, they are particularly prone to cause eye cataract.

Beneficial uses of radiation

Here we list a few out of the many applications of radiation technology which have benefited mankind. The length of the list and the importance of the items perhaps disguise the fact that the range of possible uses of these penetrating radiations is only just beginning to be realized.

(a) *Diagnostic uses in medicine and biology*: Apart from the use of X-rays, the prime example in this category is the development of *tracer* techniques, in which the metabolic paths of particular elements in an organism can be followed by replacing all or some of the natural isotope by a radioactive isotope. By using suitable detectors this isotope can be tracked in its passage through the organism. Tracer techniques have given much information, not only about pathological conditions such as diabetes, where sugar metabolism can be studied in this way, but also about the normal metabolism of both animals and plants.

Example: The path of oxygen in photosynthesis has been determined by the use of the radioactive isotope O^{18}. The process may be written as

$$H_2O + CO_2 \rightarrow (CH_2O) + O_2$$

and for some time it was unclear whether the oxygen incorporated in the carbohydrate came from the water or the carbon dioxide. By labelling the water with molecules of $H_2(O^{18})$ and noting the subsequent appearance of radioactive (CH_2O) it has been shown conclusively that the former is the case.

(b) *Therapeutic uses in medicine*: Radiation treatment is the commonest and most effective way of attacking many cancerous conditions which are characterized by rapid multiplication of the malignant cells. The principle is that although massive doses may kill normal body cells they will preferentially kill the cancerous tissue. The activity required is large, of the order of kilocuries, and is provided either by fixed installations in hospitals, such as Co^{60} sources or particle accelerators, or (for very localized treatment) by injection of a very short-lived isotope. Two specific examples may illustrate the techniques. The first is the use of radioactive I^{131} to treat hyperthyroidism. This condition had previously to be treated surgically, but by using the ability of the thyroid gland to take up ingested iodine it is possible to attack the condition in a simpler and more efficient way. The dose is administered until the excess gland cells have been killed; the remaining radioactive isotope has a short half-life and soon ceases to be a hazard.

The second example is the use of beams of charged particles from accelerators to treat brain tumours. At present protons with energies of up to 100 MeV are used for this purpose, but there is rapidly-growing interest in the use of beams of unstable pi-mesons. In either case a narrow beam is directed at the internal tumour, and radiation damage to the intervening tissue is prevented either by using particles of such an energy that they come to rest, and therefore have their largest RBE, at the site of the tumour, or by rotating the patient about an axis passing through the tumour site. In this way the surrounding tissue receives only a low dose, the area of high dose being determined by the beam dimensions.

(c) *Agricultural uses*: The propensity of radiation to produce mutation is in general harmful, simply because most mutations are themselves harmful. However, some may be beneficial, and an increased mutation rate allied with careful selection of the mutant strains has led to great advances in plant breeding. For instance, a great deal of effort has gone into producing mutant varieties of rice by intense γ-irradiation of the seeds, and this has resulted in heavy-yielding and disease-resistant strains which are now grown commercially, and with great success, in India and other major rice producing regions.

Similar genetic effects are used in pest control, specifically where insect pests with a high reproduction rate are involved. Intense irradiation is used to sterilise laboratory insects without killing them and subsequent release of numbers of sterile males into a native population can very quickly lead to reduction of the species to a level at which conventional control techniques can take over. For example, by these means the boll weevil, a cotton pest previously causing over one hundred million dollars' worth of damage annually in the USA, was virtually eliminated at a cost of only one tenth this amount.

(d) *Industrial uses*: There are many of these and their number is increasing. Two examples that may be quoted involve the use of γ-rays. In radiography, one searches for defects (fractures) in welded metals, or uses thickness gauges which rely on measuring the attenuation of the rays in the given material (which can be anything from steel plate to cigarette paper). Powerful Co^{60} sources are also used to kill bacteria, and thus sterilize packaged surgical dressings.

EXERCISES AND ANSWERS

EXERCISES

1. Certain locusts can make jumps of up to 70 cm, reaching a height of 30 cm.

Calculate the vertical and horizontal components of velocity that it is necessary to achieve on take off, in order to execute such a jump. Assuming an insect mass of 1·5 grams and that the vertical component of velocity is reached in 8 milliseconds, calculate the average vertical component of the force exerted by the insect upon the ground whilst taking off.

2. It is possible for a person to swing a bucket full of water around in a vertical circle without spilling any water. Supposing the distance from the shoulder joint to the bucket is 70 cm calculate the minimum speed at which no water will be spilled. How many revolutions per second does this correspond to?

3. A man stands with his arm held out horizontally in front of him. Is he doing any work?

4. An air-tight, perfectly efficient and well insulated domestic refrigerator, containing air at 27°C and atmospheric pressure, is switched on. After a time the interior air has cooled to 0°C. Calculate:

   (a) the force now needed to open the door.
   (b) the amount of energy used to cool the interior air.

(Assume the refrigerator to be of typical dimensions.) Comment on whether the assumptions stated in the first line are realistic. Specific heat of air at constant volume can be taken as 0·17 calories per gram per degree and the density of air as $1 \cdot 2 \times 10^{-3}$ grams per ml.

5. (a) With what velocity must a molecule, at the top of the atmosphere, travel to escape completely from the Earth's gravitational attraction?

(b) Hence find what proportion of hydrogen molecules in a gas at $1000°K$ (roughly the temperature near the top of the atmosphere) have sufficient speed to escape from the earth.

(Hint: the gravitational potential energy of a particle of mass m at a distance r from the centre of the Earth is $\dfrac{-GM_e m}{r}$ where M_e is the mass of the Earth and G is the gravitational constant.)

6. A circular pond is 1 m deep and 100 m in radius. Suppose we introduce a fish which is a bottom-feeder (and so remains always close to the bed of the pond) and which respires 1 ml of oxygen per second. Can it survive?

(Hint: find the rate at which oxygen will diffuse downwards from the surface, assuming its concentration to increase steadily from zero at the bottom to its equilibrium value of 8 ml per litre of water at the surface, all volumes being measured at STP.)

7. Some insects that remain submerged in water for long periods carry down a bubble of air with them when they dive. Initially the gases in this air store (21% oxygen, 79% nitrogen) are in equilibrium with the gases dissolved in the water (33% oxygen, 64% nitrogen, 3% carbon dioxide). Discuss qualitatively what happens as the insect uses up oxygen. (Carbon dioxide is very soluble in water so there is never very much in the bubble.)

8. The diffusion constants (D) for the diffusion of oxygen through air and through tissue are in the ratio 8×10^5. Given that the oxygen path down the tracheal system of an insect is 10^5 times longer than its path through the tissues, estimate how much longer it will take oxygen to diffuse through the tissue than down the trachea assuming that the concentration difference is the same in the two cases.

9. It is desired to use a centrifuge to separate a suspension of cells (specific gravity $1\cdot003$) from the plasma (specific gravity $1\cdot001$). What minimum RCF value is required to perform the separation? If the sample is at a distance of 4 cm from the axis of rotation, what angular velocity does this correspond to? (Assume

Exercises and Answers

the cells are 1 μm across.)

10. Using the approximate value of the activation energy for denaturation of protein, given in Table 2 2, calculate the ratio of the rate of inactivation:
   (a) at 40°C to that at 37°C.
   (b) at 0°C to that at 26·5°C.
Comment on the relevance of your answers to the lives of warm and cold blooded animals.

11. A certain ester breaks down at 50°C. The esterification reaction may proceed by a direct path or an enzyme-catalysed one. The activation energy E^* for the direct path is $\frac{176\cdot5}{\sqrt{T}}$ kilocalories, where T is the absolute temperature. E^* for the catalysed path is a constant 9·8 kilocalories, but the enzyme becomes de-activated above 35°C. Calculate and plot, as a function of T, the quantity $\frac{E^*}{T}$ at 5°C intervals between 0° and 50°C, for each path. Hence, deduce whether there is an optimum temperature for performing the reaction and if so which path should be used.

(Note: it is unnecessary to calculate $e^{-E^*/RT}$ for each value; clearly as $\frac{E^*}{T}$ goes up, $e^{-E^*/RT}$ will go down and vice versa.)

12. If a man's daily food intake is 3000 kilocalories and this exceeds the amount of energy he expends by 5%, then calculate his weight increase per year assuming he stores this as fat and glycogen (9 kcal/gram).

13. Calculate the change in free energy when 500 ml of a 0·1 molar solution of glucose in water is diluted to one litre at a temperature of 30°C (molar solution = one mole of solute per litre of solution).

14. The change in Gibbs' free energy for reactants in the standard state ($\Delta G°$) in the reaction 1, 3 diphosphoglyceric acid + ADP ⇌ ATP + 3-phosphoglyceric acid is 4050 calories at 293°K and pH = 7·8. Calculate the equilibrium constant of the reaction under these conditions.

15. The conductivity of human protoplasm is approximately $0\cdot01\,\Omega^{-1}\mathrm{cm}^{-1}$. Find the current passing through one's body if one grasps the terminals of a 2 V battery, one in each hand. Make any necessary approximation. Comment on your answer, given that 1 mA

of total body current is usually fatal.

16. The concentration of potassium ions is ten times greater in a cell than in the surrounding plasma. If the cell membrane is permeable to potassium ions, calculate the potential difference required across the membrane at 25°C so that there will be no nett flux of potassium ions across the membrane.

17. The potential difference between a hydrogen electrode immersed in a solution of hydrogen ions of concentration 0·001 gram-ions per litre and a hydrogen electrode immersed in another solution of hydrogen ions is 0·118 volts. If both solutions are at 25°C calculate the *pH* of the latter solution.

18. The human eye is a remarkably sensitive light detector. It has been shown experimentally that the average retina responds with 60% probability to about 10 photons of wavelength 510 nanometres arriving in one millisecond. (At this wavelength the rods have maximum sensitivity.) Calculate the total energy of the ten quanta. Compare your answer with the energy consumed by a 4 watt light bulb in one second.

19. The irridescence of membranous insect wings is due to destructive interference after reflection from the surfaces of a series of cuticular layers, 0·2 μm apart. Calculate a few of the wavelengths which will interfere destructively on reflection from the first and third cuticular layers. Assume that the light is vertically incident and the refractive index of wing material is 1.

20. The giant chromosomes of the fruit-fly *Drosophilia* are easily seen using an optical microscope. Can you deduce size limits from this information?

21. A suspension of spores in water is being studied under a microscope, which is operating close to its diffraction limit. The numerical aperture is one, light from a sodium lamp is used, and the spore are spheres of diameter 4·0 μm and specific gravity equal to one. It is desired to take a photomicrograph of the spores.

(a) Supposing the spores were at rest, calculate the amount of image blurring due to diffraction.

(b) In fact the spores move because of Brownian motion. If the shutter speed is 0·05 second (limited by the amount of light available) find the amount of blurring due to this effect and compare it with that due to diffraction.

22. A hawk, of typical dimension 20 cm, is hovering 300 m above the ground when it spies a mouse (dimension about 5 cm) immediately below. The hawk flies vertically downwards with accereration $2g$ until it reaches its 'terminal velocity' in air; it then continues at constant velocity until it hits the mouse.

(a) Find the blurring of the mouse image due to diffraction effects, in the hawk's eyes; assume the pupil to be 4·5 mm in diameter.

(b) Find the hawk's terminal velocity.

(Hint: use Stoke's law, given on p. 70, assuming the hawk to be a sphere of specific gravity 1. At these high speeds the air flow is turbulent and the viscosity may be taken as $0·2 \text{ kg s}^{-1} \text{ m}^{-1}$.)

(c) Hence find how long the hawk's downward dive takes. (Use the kinematic laws of Chapter 1.)

(d) Suppose that after the hawk reaches constant velocity, the mouse walks away with velocity 8 cm s^{-1}. Will the hawk catch it? (Use your judgement here.)

23. In acid solution sucrose undergoes the 'inversion' reaction sucrose → dextrose + levulose, so-called because the initial and final solutions are optically active in opposite senses. This reaction can be studied in a polarimeter by measuring the rate at which the angle of polarisation rotates, as one varies the hydrogen ion concentration of solution. The rate of inversion is found to be directly proportional to the hydrogen ion concentration.

A solution of 20 g of sucrose in 100 ml of water is poured into 100 ml of 1N HCl. A portion of the mixture is immediately poured into a 20 cm long polarimeter tube and its optical activity measured. The observed angles of rotation after 0, 10, 100, 250 and 500 minutes are $+26\frac{1}{2}°$, $+21°$, $-5\frac{1}{2}°$, $-14\frac{1}{2}°$ and $-16°$. Calculate:

(a) the *pH* of the solution.

(b) the time after which the angle of rotation has changed by one-half its total change (the half-life for inversion).

(Hint: Use log-linear graph paper.)

(c) Suppose a solution of *pH* equal to 1·3 had been used; what would the half-life have been? What rotations would have been observed after 0, 10, 100, 250, 500 minutes?

(d) Find the specific rotations of sucrose and of the final equilibrium mixture of dextrose and levulose.

24. A photographic light-meter has a special photosensitive surface with a work function of 1·5 eV. What is the maximum wavelength of light that will just emit electrons from the surface? If a scene

illuminated by sodium vapour lamps is observed by the light-meter, what voltage is required to stop all the emitted electrons?

25. A diatomic molecule has a series of vibrational energy levels whose spacing is 2×10^{-20} joule per molecule. Use the Maxwell distribution to show that the number of molecules in the first excited state at 25°C is less than 1% of those in the ground state. (This type of argument is important in making simplifying assumptions when interpreting infra-red absorption spectra.)

26. As a result of nuclear reactions induced by cosmic radiation, there is a steady level of C^{14} (half-life 5800 y) in the atmosphere, amounting to one part in 10^{-10} of all atmospheric carbon. This is incorporated into living plants, but not dead ones.

(a) A piece of wood (50% carbon) purporting to have come from an ancient Egyptian pyramid gives a C^{14} count-rate of 8 disintegrations per second per gram of wood. Is the artefact a genuine one?

(b) Peat from an Irish bog (96% carbon) gives a C^{14} count-rate of 6 disintegrations per second per gram. How old is it?

(c) Express the activity of the Irish peat in terms of curies per gram.

27. Just as electrons can be diffracted by virtue of their wave-like properties, so also can neutrons, and both can be used like X-rays to study molecular and crystalline structure.

(a) Find the wavelength of 80 000 volt X-rays.

(b) Find the energies of electrons and neutrons which have the same wavelength as the X-rays in (a). Is the voltage needed to produce this electron energy higher or lower than 80 000 V? Could the neutrons come from a reactor? Give reasons.

(c) Comment on the safety precautions that would be needed in the case of X-ray, electron and neutron diffraction studies.

28. A post-war army manual instructed soldiers that, in the event of a nuclear explosion close by, they were to 'dig a 6 ft deep trench and jump into it'. Disregarding the humourous aspect, write a detailed account of the radiation hazards against which they would, and would not, be protected by such action.

29. Suppose 1 cm of lead reduces the radiation from a hard X-ray tube to a tolerable level, say one thousandth of the unshielded intensity. The X-ray tube current, and hence the X-ray intensity, is then doubled. How much lead is required now? What is the X-ray tube voltage?

Exercises and Answers

30. Though β-rays themselves have a range of only about 1 mm of lead, one needs a much thicker layer to provide adequate radiation protection. Why?

31. Suppose that a beam of high energy protons from an accelerator can pass straight through one's body. Compare the hazards involved in (a) standing in the unobstructed path of the beam; (b) standing behind a block of lead sufficiently thick to stop the protons completely; and (c) standing behind lead of insufficient thickness to stop the protons, whose energy is thereby reduced so that they come to rest in one's body.

32. A nucleus has a constant probability λ of decaying per second. Show that the probability of its surviving for a small time interval Δt is $(1 - \lambda \Delta t)$, of surviving for two such small time intervals is $(1 - \lambda \Delta t)^2$ and of surviving for t secs is $\left(1 - \frac{\lambda t}{n}\right)^n$ where $n = \frac{t}{\Delta t}$, the number of small time intervals in t secs. Of course to get the correct answer for the probability of surviving for t seconds, n should become indefinitely large. The limit of $\left(1 - \frac{\lambda t}{n}\right)^n$ as n tends to infinity is $e^{-\lambda t}$. For the case $\lambda = 1$ sec^{-1} and $t = 2$ secs, calculate $\left(1 - \frac{\lambda t}{n}\right)^n$ for $n = 3, 8, 100$ to show that it is converging to the value of $e^{-\lambda t}$.

(Hint: If the probability of decay in Δt is p then the probability of not decaying is $(1 - p)$. If the probability of surviving the first time interval is equal to r and the probability of surviving the second time increment is q, then the probability of surviving both the first two time increments is rq. Use logarithms to do the calculation; $\log_{10}(X^n) = n \times \log_{10}(X)$.)

ANSWERS

1. To calculate initial vertical component of velocity use the conservation of energy in the form of eqn (1.36), obtaining 2·42 m s^{-1}. Then find total time in the air, from eqn (1.4), to be $2 \times 0·247 = 0·494$ seconds. Hence the (constant) horizontal component of velocity is $0·7/0·494 = 1·42$ m s^{-1}.

To find force we need mass (given) and vertical acceleration. The latter, from eqn (1.4) again, is $2·42/0·008 = 302$ m s^{-2}. Hence vertical component of force $= 0·0015 \times 302 = 0·45$ N.

Compare the height jumped: mass ratio with that of a human!

2. Requirement is that the centripetal acceleration of the bucket equal g, so that at the top of the swing the water is accelerating downwards with acceleration g. Equation (1.12) then yields velocity = $2 \cdot 62$ m s^{-1} i.e. $\omega = 2 \cdot 62/0 \cdot 7 = 3 \cdot 74$ radian s^{-1} = $0 \cdot 595$ revolutions per second.

3. Yes. Even apparently stationary muscles, when contracted, undergo continual small-scale movements, and hence do work against whatever force opposes them. A familiar example occurs when one lifts a weight above one's head; the sensation of effort disappears as soon as the elbows 'lock' and the rigid bones take the strain from the muscles.

4. Assume a $1 \times 1 \times 1$ m^3 refrigerator, with door of area 1 m^2. (a) Equation (2.4) shows that a 10% change in absolute temperature will lead to a 10% change in pressure. Initially the pressure was atmospheric, that is about 10^5 N m^{-2}, so the resultant force is finally about 10^4 N, or one tonne-weight. (b) about 2×10^4 J = $0 \cdot 006$ kilowatt-hour. (a) is an immense force, (b) is a trivial quantity of electric power, so the assumptions are unrealistic.

5. (a) By conservation of energy, and using Newton's law of gravitation, we find the 'escape velocity' to be 11 km s^{-1}.

(b) From p. 64 the proportion is about e^{-15} or one part in three million.

6. Use the approximate value for the diffusion constant of oxygen, quoted in Chapter 2. Remember that in water $\bar{v}$ remains the same but λ is reduced by a factor of about 1000. Then eqn (2.8) leads to the result that about $2 \cdot 5$ ml per second of oxygen diffuse downwards. The fish requires only about one ml per second so it survives.

7. Any CO_2 expired enters the water. Since the nitrogen : oxygen ratio is increased by the insect, nitrogen diffuses out and oxygen diffuses in. In fact oxygen diffuses in three times faster than nitrogen diffuses out. Consequently the insect can remain submerged for much longer periods than might at first be thought.

8. Use eqn (2.7): since Δn is the same in both cases

$$\frac{J_{\text{trachea}}}{J_{\text{tissue}}} = 8 \times 10^5 \times \frac{1}{10^5} = 8$$

The oxygen diffuses 8 times faster down the trachea.

Exercises and Answers

9. The RCF value must be at least 400. Putting this equal to $\frac{r\omega^2}{g}$ we find $\omega \sim 300$ radians per second or 50 revolutions per second.

10. (a) about $4\frac{1}{2}$
    (b) about 4×10^7 } using $E^* = 40 \times 10^4$ J mol^{-1}.

11. Plot $\frac{E^*}{T}$ against T for each path. The minimum value for the catalysed path occurs at 35°C; the direct path values fall below this value at 42°, so one would use the direct path, as close to 50°C as was considered safe. (If the direct path had had a *higher* $\frac{E^*}{T}$ at all temperatures, one would have used the catalysed path as close to 35°C as possible.)

12. Energy excess per year = $3000 \times 0{\cdot}05 \times 365$ kilocalories
    ∴ his weight increases by $\frac{150 \times 365}{9}$ grams = 6·1 kilograms.
    Luckily the body controls the appetite appropriately!

13. From eqn (3.24) we find:
    $$\Delta G = 0{\cdot}05 \times 1{\cdot}987 \times 303 \times \log_e \frac{0{\cdot}05}{0{\cdot}10} = 20{\cdot}8 \text{ calories.}$$

14. Substituting into eqn (3.23) gives $K \simeq 1000$.

15. Taking the cross-sectional area of one's arms as 40 cm², and the hand-to-hand distance as 2 m, one applies eqn (4.2) to find a resistance of 500 Ω. Thus by Ohm's law 4 mA will flow. This is not a dangerous exercise in fact, because the skin and various cell membranes are far worse conductors that the bulk of electrolyte-saturated protoplasm.

16. It is necessary to cancel the voltage difference set up by the concentration difference across the cell membrane.
    $$E = 0{\cdot}059 \log_{10}(10)$$
    $$= 59 \text{ millivolts}$$
    If a potential of 59 millivolts is set up in opposition to the 'concentration induced' potential there will be no nett flow.

17. Using eqn (4.16)
    $$E = 0{\cdot}059 \log_{10}\left(\frac{C}{0{\cdot}001}\right) \quad \therefore \quad \frac{0{\cdot}118}{0{\cdot}059} = \log_{10}\left(\frac{C}{0{\cdot}001}\right) = 2.$$
    ∴ $C = 0{\cdot}1$, hence $pH = -\log_{10}(0{\cdot}1) = 1$.

18. The energy of one photon = hc/λ and hence the total energy is $\dfrac{10 \times 6\cdot 6 \times 10^{-34} \times 3 \times 10^{8}}{510 \times 10^{-9}} \simeq 4 \times 10^{-18}$ joules. The light bulb consumes 4 joules.

19. In this case it is easy to show that destructive interference occurs if $2nd = (m + \frac{1}{2})\lambda$ where d is the separation of the reflecting surfaces and m is an integer (cf. p. 123, bottom; the difference between the two examples arises from a phase-change occurring at reflection). Putting $n = 1$, $d = 2 \times 0\cdot 2 = 0\cdot 4 \mu$m, we find $\lambda = 16\,000\,\text{Å}$, $5333\,\text{Å}$, $3200\,\text{Å}$ for $m = 0, 1, 2$ respectively.

20. You can deduce only that the chromosomes are larger than the wavelength of light in the visible region — say larger than 500 nanometers.

21. (a) Equation (5.7) tells us that the Airy disc of a point source has a radius of about $0\cdot 36 \mu$m (using $\lambda \sim 600$ nm). Thus no detail smaller than about $\dfrac{0\cdot 36}{4} \sim \dfrac{1}{10}$ of the size of the spore can be resolve

(b) Equation (2.2) tells us that the spore's mean velocity is $\sim 0\cdot 6 \times 10^{-3}$ m s^{-1}, and substituting in eqn (2.5) we find the distance travelled in $0\cdot 05$ s to be about $0\cdot 2 \mu$m. (Using 10^{-9} m as a rough valu for λ, the mean free path; take care over these two uses of the symbol λ.) The two effects are therefore similar in magnitude.

22. (a) From eqn (5.5a) the angle θ of the cone into which the light is diffracted is $\dfrac{\lambda}{W}$ where $\lambda \sim 450$ nm and $W = 4\cdot 5$ mm. Hence $\theta = 10^-$ radians corresponding to a blurring of 3 cm at 300 m.

(b) Approximately 100 m s^{-1} ($\approx$ 200 mph; this great speed is reached by some falcons).

(c) Using eqn (1.4) we find the hawk accelerates for 5 s and flie at steady speed for $\frac{1}{2}$ s, the total dive-time being $5\frac{1}{2}$ s.

(d) In $\frac{1}{2}$ s the mouse travels 4 cm. The hawk would probably still catch it.

23. (a) The *pH* is $0\cdot 301$ (eqn 4.15).

(b) 50 minutes.

(c) 500 minutes (since the hydrogen ion concentration is ten times smaller, the rate of inversion is ten times less); $+ 26\frac{1}{2}°$, $+25\cdot 9°$, $+21°$, $+14°$, $+5\frac{1}{4}°$.

(d) $+66\cdot 4°$; $-40\cdot 0°$.

Exercises and Answers 217

24. From eqn (6.1) we find 825 nm light will just emit an electron with zero kinetic energy. Sodium light has $\lambda = 589$ nm, and the voltage needed is 0·6 V.

25. $$\frac{\text{No. in first excited state}}{\text{No. in ground state}} = e^{-\frac{2 \times 10^{-20}}{298 \times 1\cdot38 \times 10^{-23}}}$$
$$= 0\cdot008.$$

26. (a) No. The steady level of C^{14} yields 19·2 disintegrations per second (dps) per gram of carbon. The recorded level of $(8 \times 2) = 16$ dps per gram corresponds to an age of $\frac{5800}{0\cdot693} \log_e \left(\frac{19\cdot2}{16}\right) \approx 1500$ years (from eqn 7.3). This is not even pre-Christian.
  (b) About 10 000 years.
  (c) $1\cdot62 \times 10^{-10}$ curies per gram.

27. (a) Using eqn (6.2), $\nu = E/h = \frac{80\,000 \times 1\cdot6 \times 10^{-19}}{6\cdot63 \times 10^{-34}}$ $1\cdot93 \times 10^{19}$ s^{-1}. Hence $\lambda = \frac{c}{\nu} = 1\cdot55 \times 10^{-11}$ m.
  (b) From eqn (6.4), we find the electron energy to be $1\cdot07 \times 10^{-15}$ J ≈ 6700 eV. Hence only 6700 V is needed to produce this energy. Since neutrons are about 1840 times more massive than electrons, their energy for a given wavelength is 1840 times less, or about 3·6 eV. Since the energy of fission neutrons is typically 1 MeV, they could be slowed up ('moderated') to this energy.

29. About 1·1 cm. Approximately 250 kV (see Fig. 7.4).

30. Because the bremsstrahlung produced are γ-rays and hence very penetrating.

31. (b) is safest, followed by (a) and (c) in that order. In (c) the RBE of the slow protons near the end of their range is very high.

32. $\qquad\qquad e^{-2} = 0\cdot135$

$\left(1 - \frac{2}{n}\right)^n$	n
0·037	3
0·100	8
0·133	100

BIBLIOGRAPHY

Quite possibly some readers of this book (we hope many!) will want to carry their study of physics further, in which case they will need a 'physicist's' textbook. Of the enormous number available we have selected one in particular as being lucid, relevant, not too mathematical and written in a superbly elegant and witty style. It is *Lectures on Physics* by R.P. Feynman, R.B. Leighton and M. Sands (Addison-Wesley, 1963). Volume I deals with most of the topics we have covered here.

We recommend two books dealing with the basic biophysical ideas of biology. They are *Plants at work* by F.C. Stewart (Addison-Wesley, 1964) and *Animal mechanics* by R. McN. Alexander (Sidgwick and Jackson, 1968). They show how the ideas we have developed here are applied in the fields of botany and zoology respectively. Finally, for those who wish to supplement the exercises of this book, we suggest they try those in *Examples in quantitative zoology* by M. Jarman (Arnold, 1970), a compilation of nearly 80 questions with solutions.

In addition to these general references, the following books and articles will amplify either whole chapters of this book, or sections of chapters.

Chapter 1: Mechanics

A programmed introduction to vectors, R.A. Carman (Wiley, 1963). A teach-yourself text for all levels of ability.

Foundations of physics, R.B. Lindsay and H. Margenau (Dover Books, 1957). See ch. 3 for a clear account of the concepts of mass, force and coordinate systems.

The meaning of relativity, A. Einstein (Methuen, 6th ed., 1967). Eminently readable introduction for the layman.

Animal locomotion, Sir James Gray (Weidenfield and Nicolson, 1968). A complete and lavish textbook by the leading authority.

Chapter 2: The Molecular Nature of Matter

Gases, liquids and solids, D. Tabor (Penguin Library of Physical Sciences, 1969). Discusses the topics of this chapter with great clarity — see especially ch. 6 for deviations from the ideal gas laws in real gases.

Reverse osmosis. There is no book yet on this new subject. See the articles by A. Sharples (*Science Journal*, August 1966), W.H. Hardwick (*New Scientist*, 9 April 1970) and O. Illner-Paine (*Science Journal*, December 1970).

Movement of sap in trees. Again, for up-to-date ideas on this topic, see the articles by M.H. Zimmerman (*Scientific American*, March 1963) and A.T. Hayward (*New Scientist*, 29 January 1970).

The control of biological reactions, J.-P. Changeux (*Scientific American*, April 1965). How enzymes control reaction rates.

Chapter 3: Thermodynamics

Bioenergetics, A.L. Lehninger (Benjamin, 1965). Gives a full account of the biological applications of thermodynamics, which we could mention only briefly.

Chapter 4: Electrochemistry

Modern physical chemistry, M.F.C. Ladd and W.H. Lee (Penguin Library of Physical Sciences, 1969). Covers most of the material of this chapter in much greater detail.

The nerve axon, P.F. Baker (*Scientific American*, March 1966). Perhaps the best of many recent articles on the measurement of membrane potentials. Though lacking the detail of a textbook treatment it gives a lucid account of a very complex subject.

Chapter 5: Optics and the Wave Nature of Light

What is light?, A.C.S. van Heel and C.H.C. Velzel (Weidenfeld and Nicolson, World University Library, 1968). Copiously illustrated, chatty account. Strongly recommended as a follow-up to this chapter; particularly good on lasers.

From sight to light, R.A. Weale (Oliver and Boyd, 1968). The physiology of vision explained in relation to the physical nature of light. Many amusing and paradoxical examples, both visual and verbal.

Chapter 6: Modern Physics: Particles, Waves and Probabilities

The double helix, J.D. Watson (Weidenfeld and Nicolson, 1968; reprinted Penguin, 1970). Fascinating account of how the structure of DNA was discovered. As good for its insight into the conduct of research as for its science.

The scanning electron microscope, J.A. Clarke and others (*Science Journal*, August 1968). Describes what is perhaps the most significant advance in microscopy for decades.

Organic semiconductors, D.D. Eley (*Science Journal*, December 1967). Many examples chosen from biophysics.

Solid state biophysics, Ed. S.J. Wyard (McGraw-Hill, 1969). A series of review articles describing how the techniques of electron spin resonance, lasers, etc. are applied in medical and biological science.

Chapter 7: Nuclear Radiations

Introduction to health physics by H. Cember (Pergamon Press, 1969). Readable, thorough and expensive, with a good account of the physics involved.

Sunburn, Farrington Daniels Jr. and others (*Scientific American*, July 1968). Discusses what we know of a common, though complex, example of radiation damage.

Genetic effects of radiation, C.E. Purdam (Newnes, 1963). Still the clearest account of genetic damage.

INDEX

aberrations, *see* lens
absorption coefficient, 185—191
accelerating definition, 6
  in SHM, 27—29
activation energy, *see* energy
activity, 98—9
Airy disc, 132—3
alpha particle, 25, 179—180, 188—189
ampere, 90
amplitude (of vibrations and waves), 31
angular velocity, 8, 29
  momentum, 13, 166—7
anode, 90, 93 (note), 150
Apollo project, 127, 156
armalcolite, 156
atmosphere, pressure variation, 58—59
atomic number, 151, 174
  mass unit, 174—7
attenuation coefficient, 185—191
Avogadro's number, 41, 92

Babinet's theorem, 127
Balmer series, 162—3
Becquerel, 178
beta particles, 179—182, 189, 191
binding energy, *see* energy
birefringence, 142
Boltzmann's constant, 40, 74, 148
Born, 164
Bohr, 166
Boyle's law, 41
Bragg, 154—6
  equation, 156—7

bremsstrahlung, 188, 190 (table)
Brewster's angle, 139
Brownian motion, 42—45, 49, 51

calomel reference electrode, *see* electrode
capillary flow, 71
Carnot, 82—3
catalysis, 68
cathode, 90, 93 (note), 150
  rays, 150
cell, *see* electrochemical cell
centrifugal force, 11, 60
centrifuge, 60
centripetal acceleration, 9—10, 11, 16, 24
chain reaction, 178
Charles' law, 41
chemical reactions, *see* reactions
circular motion, 6—10, 161
cluster formation, 92—3, 97
coherence, 123, 143—6
Compton scattering, 185—6, 190 (table)
concentration cell, 103—7
conductance, equivalent, 92—4
conduction band, 169—70
conductivity, 91
conductor, electrical, 169—70
Coolidge tube, 154
cosmic rays, 184, 192
coulomb, 90
counter, 194
critical mass, 178
Curie, 178
curie (unit), 183

221

Dalton, 149
Dalton's law of partial pressures, 41, 54
Daniell cell, 100–3
Davisson, 156–7
de Broglie, 157, 164
degrees of freedom, 42
determinism, 148
diffraction, 124–136
  grating, 128–130
  of electrons, 156–157
  of X-rays, 154–156
diffusion, 45, 47–52, 57
  constant, 47, 49, 51
  example of insect respiration, 49–50
  Graham's law of, 50
  of water, 52–7
  potential, 105
dissociation, ionic, 92, 107
dosimetry, see radiation dosimetry
duality, 157, 161
dynamics, 10–15

efficiency, of heat engine, 82–3
Einstein, 12 (note), 25, 43, 148, 175
electric current, 89–91, 95
  field, 23–24, 91, 95
  potential, 89–90, 98–9, 102–3
e.m.f., 90, 99, 102–3
electrochemical cell, 98–107
electrode, calomel, 99
  hydrogen, 99, 106–7
  potential (reversible), 98–107
electrolyte, 91
electrolytic bridge (salt bridge), 100, 102, 104–5
  conduction, 91–8
electron, 89–90, 150, 174
  spin resonance (esr), 168
  wave nature of, 156–60
electron volt, 22, 152
electroscope, 192
energy, 20–24
  activation, 67, 68 (table)
  binding, 163, 176–80
  conservation of, 24–26, 73–75
  electrical, 23–24
  equipartion of, 42–43
  internal, 74, 84, 85 (table)
  kinetic, 20–1
  levels, 160–4, 166–71, 182
  of molecular motion, 39–40, 42
  potential, 21–3
  of vibrating spring, 27–9
enthalpy, 76–80, 85 (table)
  of standard state, 79 (table)
entropy, 80–4, 85 (table)
enzymes, 68, 78–9
equilibrium, definition of, 43 (note)
  thermal, 42, 84
  sedimentation, 59
  constant, 67, 85
  chemical, 65–7
equivalent conductance, see conductance
equivalence, chemical, 92
eriometer, 128
evaporation, 61–63
exclusion principle, 168–9
eyes, compound, 134–6
  sensitivity of, 153

fall-out, 182
Faraday, 92
Fermat's principle, 112–115
Fick's Law, 47, 86
film badge, 195
fission, 177–8, 182
  fragment, 178, 200
flow, capillary, 71
  streamline, 71
  turbulent, 71
force, 10, 14–15
  centrifugal, 11, 60
  electrical, 23–24
  of stretched spring, 29, 31
  gravitational, see gravity
'fountain pen' dosimeter, 192, 196 (table)
Franck, 163–4
Fraunhöfer lines, 162 (note)
free fall, 11
frequency, 33, 111
friction, 11, 13, 15, 69, 82
fundamental constants, 148
fusion, 177

Galileo, 31
gamma rays, 179, 182, 186–9 199–201
gas law, 41, 54–5
  constant (R), 41
  ideal, 41, 54–5, 74, 93
  discharges, 149–151
Geiger–Müller counter, 193–7

Index

geometrical optics, 112–117
Germer, 156–7
Gibbs' free energy, 83–7, 85 (table), 101, 103
glancing angle, definition of, 155
Graham's law of diffusion, 50
gravity, 10, 15–17, 57–8
ground state, 163

half-cell, 99–106
half-life, 183
health physics, 173
heat, 40, 74–5, 76–80, 82–3
hertz (unit), 111
Hertz, 151, 163–4
hole, 170
holography, 146
Hooke's law, 29

ideal atmosphere, 58–9, 64
ionic velocity (mobility), 94–7
ionization, 150, 163, 185–205
ions, 55, 91–98, 150
insects, compound eyes of, 134–136
insulator, electrical, 169
interference, of light, 117–23
  fringes, 121–7
  of X-rays, 154–6
  of electrons, 156–7
isothermal, definition of, 58
isotope, 175–8, 182

Joule, 37
  (unit of energy), 19

kinematics, 5–10
  of circular motion, 6–10
kinetic theory, 37–42, 61, 69

laser light, 127, 143–6
  uses of, 145–6
Lavoisier, 40
lenses, 112, 115–6, 131–4
  aberrations of, 116, 132, 134
LET (linear energy transfer), 199
light, circularly polarised, 137–8
  plane polarised, 111, 136–8
  unpolarised, 138
light sources, 111–112
light-meter, 153
Lyman series, 162

mass attenuation coefficient, 186
mass, definition of, 14 (note)

mass defect, 175
matter waves, 157–160, 164–5
Maxwell, 37
  distribution, 63–8, 80
mean free path, 44, 45–47, 51, 54
mean life, 183
  range, 189
membranes, 51–7, 86–7, 105–6
  semi-permeable, 53–7
meson, 184
microscope, optical, 130–4
  electron, 131, 159–60
  polarizing, 142
microwaves, 167
mobility, *see* ionic velocity
moderation of neutrons, 196–7
molecules, 37–72
molecular mean free path, 45–47, 51, 54
  size, 46
  velocity, 38–40, 47, 54, 63–65
momentum, angular, 13
  conservation of, 12, 26, 35
  definition of, 11
monochromatic light, 120, 143–144
  definition of, 112

nerve impulses, 105–6
neutrino, 25, 180–181, 191 (note), 197
neutron, 174–8, 189–191
  thermal, 191
  detection of, 195–6
Newton's laws of motion, 10–15, 38
  corpuscular theory of light, 109–10
  law of gravitation, 15–17
  (unit of force), definition of, 14
Newtonian fluid, 72
nucleon, 173, 174 (note)
nucleus, atomic, 15, 25, 150–1, 173–205
numerical aperture, 132, 160

Ohm's law, 90–1, 95
oil immersion, 134
ommatidia, 134–6
optical activity, 142–3
optic axis, 141–2
orbital quantum number, 166–7
orbitals, 167
osmosis, 52–7, 86–8
  defintion of, 53
  reverse, 56
osmotic pressure, 53–7, 93

pair production, 186, 190 (table)
paraxial rays, 115–6
partial pressure, 41, 54
Pauli, 168
　exclusion principle, 168–9
period (of vibrations and waves), 31
permeability, 51–2, 53 (table)
permittivity, 23
pH, 106–7
phase of vibrations and waves, 31, 33, 118–130
　difference, 34
photoelectric effect, 151–3, 186, 190 (table)
photoelectrons, 151
photographic emulsion, 195
photomultiplier, 153, 195
photosynthesis, 153, 203
photon, 112, 144, 151–3, 162
Planck's constant, 149, 152, 157, 166
plants, evaporation in, 63
　osmosis in, 57
Poiseuille's law, 71
polarization of light, 111, 118, 136–143
　by transmission, 138–9
　by reflection, 139–140
　by scattering, 140–1
polaroid, 138–140
positron, 188
potential, electric, see electric potential
　electrode, see electrodes
　energy, see energy
　liquid junction, 105
power, 19–20, 34–35
pressure, definition of, 19
　of a gas, 19–20, 37–41
principle quantum number, 162, 166
proton, 174–8
Proust, 149

quantization, 160–4, 166–8
quantum, effects, 151
　numbers, 162, 166–8
　physics, 149

rad, 199–200
radian, definition of, 7 (note)
radiation damage, 197–203
　dosimetry, 198–203
radical, 198

radioactivity, 178–185
radioactive tracers, 203
range, 189
Raoult's law, 86–7
RBE (relative biological effectiveness), 200, 203
RCF (relative centrifugal force), 60
reaction, equilibrium, 67, 84–5
　kinetics of, 65–8, 77–9, 83–8
　rates, 67–8
rectifier, 170
refractive index, 111
　ordinary and extraordinary, 141–2
relativity, theory of, 12 (note), 25
rem, 200–202
resistance, 90–1
resistivity, 90–1
resolving power, 130–136, 159–160
　definition of, 130–1
reverse osmosis, 56
reversible electrode potential, see electrode
　process, 81–3, 98, 101
Reynolds number, 71–2
Roentgen, 153
roentgen (unit), 199
Rumford, 37
Rydberg constant, 162–3

saccharimeter, 143
salt bridge, see electrolytic bridge
satellite, motion of, 16–17, 24
scaler, 194–5
scalar quantities, 1
Schrodinger, 164
　equation, 164–6
scintillation counters, 195, 196 (table)
sedimentation, 57–61
semiconductors, 169–171
simple harmonic motion (SHM), 27–36, 118–120
semi-permeable membranes (SPM), 53–7
series limit, 163
Snell's law, 113–6
sound waves, 32
specific rotation, 143
spectrometer, 129
　Bragg, 156
speed, 6
spin, electrons, 167

Index

standard state, 80 (table), 86
 electrode potential. 101, 102 (table), 103
Stoke's law, 70
stopping power, 189
streamline flow, 71
sunburn, 198

temperature, definition of, 40
 absolute (kelvin), 40
 in thermodynamics, 74—5
thermodynamics, 73—88
 first law of, 73—6
 second law of, 81—4
tracers, *see* radioactive tracers
transistors, 170
transpiration, 23, 57
transport number, 94—97
turbulent, flow, 71

ultracentrifuge, 60
uncertainty principle, 158—160, 164

valence band, 169—170
vector quantities, 1—5, 111
 definition, 2
 components, 4
velocity, 5—6
 circular motion, *see* angular velocity

 light, 110
vibrational motion, 27—34
 of molecules, 41—42
viscosity, 68—72
 units of measurement, 69
 coefficient of, 69
 table of, 70
 specific, 70
 measurement of, 70
 of blood, 72
volt, 90

Watt, unit of power, 19
wave motion, 32—34, 109—112
wave trains, 111—2, 122, 144
wavelength, 33
waves, electromagnetic, 110
 longitudinal, 32
 transverse, 32, 111
work, 17—20, 74—5, 76—7, 82—4
 function, 152—3

X-rays, 153—6, 185
 characteristic, 163
 hard and soft, 154
 diffraction of, 154—6, 159

Young, 110